NMR CASE STUDIES

NMR CASE STUDIES

Data Analysis of Complicated Molecules

Jeffrey H. Simpson

ELSEVIER

Elsevier
Radarweg 29, PO Box 211, 1000 AE Amsterdam, Netherlands
The Boulevard, Langford Lane, Kidlington, Oxford OX5 1GB, United Kingdom
50 Hampshire Street, 5th Floor, Cambridge, MA 02139, United States

Notices
Knowledge and best practice in this field are constantly changing. As new research and experience broaden our understanding, changes in research methods, professional practices, or medical treatment may become necessary.

Practitioners and researchers must always rely on their own experience and knowledge in evaluating and using any information, methods, compounds, or experiments described herein. In using such information or methods they should be mindful of their own safety and the safety of others, including parties for whom they have a professional responsibility.

To the fullest extent of the law, neither the Publisher nor the authors, contributors, or editors, assume any liability for any injury and/or damage to persons or property as a matter of products liability, negligence or otherwise, or from any use or operation of any methods, products, instructions, or ideas contained in the material herein.

Library of Congress Cataloging-in-Publication Data
A catalog record for this book is available from the Library of Congress

British Library Cataloguing-in-Publication Data
A catalogue record for this book is available from the British Library

ISBN: 978-0-12-803342-5

For information on all Elsevier publications
visit our website at https://www.elsevier.com/books-and-journals

Working together
to grow libraries in
developing countries

www.elsevier.com • www.bookaid.org

Publisher: John Fedor
Acquisition Editor: Kathryn Morrissey
Editorial Project Manager: Anneka Hess
Production Project Manager: Paul Prasad Chandramohan
Cover Designer: Mark Rogers

Typeset by SPi Global, India

Contents

1

Introduction to the Methods of Nuclear Magnetic Resonance

This book focuses on the interactions of nuclear spins in carbon-containing (organic) compounds. The data contained in this book is derived from observing the nuclear magnetic resonance (NMR) signal of two isotopes: ^{1}H and ^{13}C, although ^{2}H and ^{19}F generate signal splittings we observe in some of the ^{1}H and ^{13}C spectra. Despite this apparent limitation in scope, it will be seen that these two nuclides furnish copious quantities of information that can be correlated with physical attributes.

NMR spectroscopy is a method rife with dogma and jargon. It would be foolish to attempt to provide a stand-alone treatise here to allow a neophyte to progress to the level of sophistication commensurate with the demands of the material presented in this work. It is suggested that the reader familiarize him- or herself with the basics of ^{1}H and ^{13}C NMR studies of simple organic molecules *before* reading this book. The remainder of this chapter provides a summary of concepts implicit in subsequent discussions. If any of the cursory descriptions are still unclear following a careful reading, the reader may wish to seek more detailed information from other sources.

1.1 NUCLEAR SPIN

An atom consists of two parts: (1) the less massive, diffuse, negatively charged electron cloud, and (2) the massive, high density, well-localized, positively charged nucleus. NMR spectroscopy addresses the interaction of low energy photons with nuclei.

The small size of the nucleus combined with a nonuniform distribution of positive charge puts nuclei into a sufficiently small scale such that the Heisenberg uncertainty principle applies: One cannot simultaneously know particle position and momentum with arbitrarily fine precision. We observe that nuclear charge is, for NMR-active nuclei, constantly in motion with a zero point energy. Nuclear charge motion is not a random walk as we envision for the electrons in the atomic electron cloud, but is circular—giving rise to a magnetic field owing to a circulating charge as required by Maxwell's equations governing electromagnetism.

NMR Case Studies
https://doi.org/10.1016/B978-0-12-803342-5.00001-6

Because the circular movement of the positive charge in the NMR-active nucleus requires physical rotation of the nucleus itself, we find that movement of the nucleus is best described using the units of angular momentum, which is rotation of a certain mass with a certain radius per unit time. We refer to a rotating nucleus as a spin.

The small size of the nucleus has a second consequence that causes the NMR-active nucleus to behave in a manner inconsistent with our normal macroscopic life experiences. Quantum mechanical expressions describe how nuclear angular momentum varies.

The quantum mechanical spin states available to a given nucleus depend on which isotope of which element we elect to study. ^{1}H and ^{13}C are both spin-1/2 nuclei (spin quantum number $I = 1/2$) and so only have $2I + 1$ (two) allowed spin states.

The energy gap between allowed spin states increases linearly with applied magnetic field strength. The energy gap for protons (hydrogen-1) in a sample we place in a strong magnetic field (inside our NMR magnet) will double if we double the strength of the applied field.

1.2 AN ENSEMBLE OF SPINS

If we place a large ensemble of spin-1/2 spins into a constant field, we can use the Boltzmann equation to calculate how the spins will be distributed between the lower and upper spin states. Under normal conditions, only one spin out of every 10^4 spins will reside in a lower energy versus an upper energy spin state.

Partitioning of spins between the lower and upper spin states only informs us of the component of the nuclear magnetic moment along the axis of the applied magnetic field (the z-axis). Exactly half of the magnetic moment of a nuclear spin will be either parallel or antiparallel to the direction of the applied field. The other component of the magnetic moment will be somewhere in the horizontal (xy) plane. The horizontal components of spins in an ensemble are, at equilibrium, evenly distributed such that the vector sum of these xy components is zero.

If we sum up the z-components of an ensemble of spins, we find that the more populated lower energy spin state "wins" the cancelation battle and the resulting magnetic vector (the net magnetization vector) points exactly parallel to the axis of the applied field. By applying electromagnetic radiation with the correct frequency, we can change the direction of the net magnetization vector.

1.3 DETECTION OF THE NMR SIGNAL

If the net magnetization vector has an xy component, it will precess (just as the rotation axis of an individual nuclear spin precesses in an applied field) and induce an electrical current in the receiver coil of our NMR probe. By capturing the amplitude, frequency, and phase of this signal, we are able to extract information pertaining to the sample that generated the signal.

1.4 INFORMATION CONTAINED IN THE NMR SIGNAL

The information available using NMR spectroscopy is all contained in the amplitude, frequency, and phase information in the superimposed sinusoidal signals that comprise the free induction decay (FID). Conversion of the time-domain signal we detect to the frequency domain spectrum we interpret is accomplished with a Fourier transform or two. The NMR spectrum is a histogram of intensity versus frequency.

For a one-dimensional (1-D) spectrum, the format is typically that on the y-axis the intensity appears as a function of a frequency range on the x-axis. The total frequency range we observe is the spectral window. The width of the spectral window is measured in Hz, and any point in the spectrum is said to have an offset from a zero point agreed upon using the NMR signal of a chemical standard. We rarely show frequency numbers on our x-axis, instead using parts-per-million (ppm). The exception is when we discuss coupling constants which are invariant with respect to applied field strength.

To obtain ppm, we determine the offset of our x-axis value from that of the standard and divide that value (in Hz) by the operating frequency of the nuclide being observed (also in Hz) and multiply the result by 10^6. We multiply by 10^6 to bring the magnitude of the x-axis numbers to within two orders of magnitude from unity for most systems. That is, NMR involves signals that differ, relative to the principal interaction (Larmor) frequency, fractionally in a range spanning between 0.0001 and 0.00000001, or in percentage terms, between 0.01% and 0.000001%, or, using the ppm standard, between 0.01 and 100 ppm. This side-by-side comparison helps illustrate why the ppm scale is so widely embraced in the NMR community.

For a two-dimensional (2-D) spectrum, the format shows intensity using contour lines or shading, with ppm along both axes. A 2-D spectrum will often resemble a topographical map, which shows altitudes as a function of longitude and latitude.

1.4.1 Chemical Shift

Chemical environment influences the energy gap between allowed states of nuclear spins. This is caused by the diamagnetic resistance of the electron cloud to the dense crowding of magnetic field lines in the sample. Because different functional groups affect electron density at and/or near the nucleus, signal frequency (chemical shift) correlates with chemical environment.

A spectrum will typically contain a number of distinct signals or resonances. Signal amplitude will, under the right conditions, correlate with spin abundance. Varying signal intensity can be used to deduce spin abundance at specific molecular sites, each with its own chemical environment. We can measure peak areas (integrals) and easily calculate relative stoichiometry of the various NMR-active sites in our solute molecule.

Most of the time the phase of the observed signal is only of concern insofar as correct phase balancing of the two orthogonal receiver channels (x and y) is or is not attainable (most spectra show only signals that are fully absorptive, with no dispersive character). However, many NMR methods are more involved than simple 1-D experiments. Observation of how the phase character of a particular signal evolves as a function of one or more preacquisition pulses (*rf* and field gradient) and delays provides detailed and specific information.

1.4.2 Intensity

Signal intensity reflects spin abundance if all detected spins of the same nuclide being compared are fully relaxed (relaxation delay > five times the spin-lattice relaxation time), are excited equally (no pulse rolloff, no nuclear Overhauser effect (NOE)), and are detected with equivalent sensitivity (no signals are attenuated due to filter edge proximity). If we attempt to compare the signals of different nuclei, we must account for Boltzmann statistics, isotopic abundance, and the magnitude of the gyromagnetic ratio.

1.4.3 Spin-Spin Coupling

NMR-active spins in our sample interact with one another through chemical bonds (the molecular electron cloud). Through-bond interactions of spins is called spin-spin coupling (J-coupling). Multiple spins are said to comprise a single spin system if every spin in the system is coupled to at least one other member of the spin system and it is impossible to further subdivide the spin system into two smaller spin systems without having intersystem coupling. When two spins are coupled to one another, we denote with a leading superscript the number of chemical bonds separating the two spins, and we may also denote the identity of the two coupled spins with a trailing subscript. For example, a $^1J_{CF}$ means we are discussing the coupling of ^{19}F directly attached to (one bond away from) a ^{13}C.

1.4.4 Dipolar Coupling

NMR-active spins also interact with one another through space. Through-space interactions of spins arise from the dipolar interaction and are termed the NOE. When we consider the dipolar interaction, we think of the ^1H's as being abundant, and of the ^{13}C's as being dilute.

1.4.5 NMR Experiments Commonly Used

A number of NMR experiments are available to exploit the through-bond and through-space interactions of spins (Table 1.4.5.1). Depending on the information one wishes to obtain, one can choose one or more of the available NMR experiments to provide this data.

A number of 1-D experiments analogous to the earlier 2-D counterparts can also be carried out. The advantage of the 1-D NOESY or TOCSY experiment is that the time required to achieve comparable sensitivity (a given signal-to-noise ratio) is considerably shorter than for the 2-D NOESY or TOCSY experiment, but the disadvantage is that by only investigating particular regions of the spectrum we may miss some other important spectral feature whose appearance was unanticipated. Only by collecting a full 2-D spectrum will we observe all of the cross peaks generated by a particular technique using a particular NMR experiment. Sometimes the most interesting discoveries occur by accident—if we do not take the trouble to look in places where our expectation of finding something interesting is low, we will surely find fewer pearls.

TABLE 1.4.5.1   Two-Dimensional NMR Experiments Commonly Used

Experiment Name	Pseudonyms, Aliases, Equivalent Experiments	Information Content	Approximate Experiment Time for a 20 mM Sample
Correlation spectroscopy (COSY)	gCOSY, avCOSY, absolute value COSY, COSY45	Homonuclear, through-bond effect; generates absorptive cross peaks between directly coupled spins (through bond)	Short (3–12 min)
Phase-sensitive correlation spectroscopy	COSYps, COSY90, DQCOSY, DQFCOSY	Homonuclear, through-bond effect; generates mixed-phase cross peaks between directly coupled spins. Cross peak phase is modulated by active coupling (coupling generates cross peak)	Intermediate to long (20–120 min)
Total correlation spectroscopy (TOCSY)	HOHAHA, homonuclear Hartmann-Hahn	Homonuclear through-bond effect; generates cross peaks between spins in the same spin system	Intermediate (12–30 min)
One-bond heteronuclear correlation	HMQC, HSQC, deHSQC, gHSQC	Heteronuclear through-bond effect; generates cross peaks between ^{1}H's and the ^{13}C's to which they are directly attached	Intermediate (12–30 min)
Multiple-bond heteronuclear correlation	HMBC, gHMBC, COLOC	Heteronuclear through-bond effect; generates cross peaks between ^{1}H's and ^{13}C's two to five bonds distant	Long (50–120 min)
Through-space homonuclear cross relaxation	NOESY	Homonuclear, through-space effect; generates cross peaks between ^{1}H's fewer than 0.5 nm apart; certain sample mobilities (if FW ~ 1 kDa) may render experiment insensitive	Long (50–120 min)
Through-space homonuclear cross relaxation	ROESY (rotating-frame Overhauser effect spectroscopy)	Homonuclear, through-space effect; generates cross peaks between ^{1}H's fewer than 0.5 nm apart; less sensitive than NOESY, but often works when NOESY fails	Long (90–180 min)

1.5 SUMMARY OF METHODOLOGY USED TO PAIR SITES AND SIGNALS

The NMR-active spins in the molecules we study generate the NMR signals we observe. Each observed resonance can be paired with a molecular site. A resonance contains one or more Lorentzian lines. The discovery process by which we pair sites and signals is assignment. We assign sites to signals and signals to sites. We can do this by annotating a picture of our molecule and writing in the chemical shifts near the sites, or by annotating the spectrum with the IUPAC number corresponding to every NMR-active molecular site, or both. Using careful accounting, we can pair every site with every signal leaving no orphaned sites nor signals.

We begin an assignment by finding entry points. Entry points are NMR signals that we can pair to a molecular site with certainty. An entry point may be the ^{13}C resonance of a carbonyl site, the ^{1}H signal of a methyl group, or a cross peak well removed from others in a given 2-D spectrum. Entry points can be found based on chemical shift, signal intensity, and resonance splittings. We use all possible means to identify entry points. Redundancy is desired.

There are five spectra we often use for the NMR assignment of organic molecules: the 1-D ^{1}H NMR spectrum, the 1-D ^{13}C NMR spectrum, the 2-D ^{1}H-^{1}H COSY NMR spectrum, the 2-D ^{1}H-^{13}C HSQC NMR spectrum, and the 2-D ^{1}H-^{13}C gHMBC NMR spectrum.

The 1-D ^{1}H NMR spectrum contains signals that allow the extraction of three types of information: (1) signal position (chemical shift) that correlates with chemical environment, (2) resonance splittings (multiplets) that correlate with spin-spin coupling through the electron cloud, and (3) signal intensities (integrals) that reflect spin abundance. We measure the position of a given resonance as the intensity midpoint, meaning that the chemical shift of a resonance is not the midpoint between its extrema on the shift scale but rather the point on the shift scale where 50% of its total integrated intensity is found. Resonances are split by other nearby spins because the states of nearby spins perturb the electron cloud. Once we identify a particular resonance, we measure the differences between the lines or legs that comprise the multiplet that is the resonance. Using statistics and a knowledge of how bond geometry affects spin-spin coupling, we can derive the magnitudes of various couplings between one spin and multiple other spins. Through the process of integration, we divide our ^{1}H signal intensity into discrete regions and normalize the measured intensities to obtain integral values correlating with ^{1}H abundance or stoichiometry at various molecular sites.

The 1-D ^{13}C NMR spectrum contains less information than the 1-D ^{1}H NMR spectrum. The 1-D ^{13}C NMR spectrum provides us with signals from which we can obtain chemical shift information and more qualitative (nonstochiometric) intensity information. If the ^{13}C nucleus generated a stronger signal, we might observe it in a ^{1}H-coupled condition, which, for example, would cause the ^{13}C signal from a methyl group to be split into a quartet. But the ^{13}C signal is weak compared with that of ^{1}H. The receptivity of the ^{13}C is lower than that of ^{1}H owing to a lower gyromagnetic ratio (a factor of between 16 and 64) *and* a lower natural abundance (1.1% vs 100%). With ^{1}H-decoupling taking place during relaxation, excitation, and detection, dipolar cross relaxation (the NOE) enhances the ^{13}C signal by as much as 200% while collapsing into singlets the multiplets of ^{13}C's with ^{1}H's directly attached. Under normal ^{13}C experimental conditions, protonated carbon sites generate stronger signals while the non-pronated sites generate weaker ones. The intensity contrast is often stark, for nonprotonated ^{13}C's have no nearby ^{1}H's from which to receive an NOE and also relax slowly enough relative to the typical scan repetition rate such that only the first of a 64 transient signal averaging receives a stochiometrically commensurate contribution, with the other 63 scans receiving the diminished contribution from a population of ^{13}C's unable to relax back to equilibrium before the next *rf* pulse occurs.

The 2-D ^{1}H-^{1}H COSY NMR spectrum is a homonuclear NMR experiment. The COSY spectrum will contain ^{1}H chemical shift axes along both axes. Because the COSY spectrum is a homonuclear experiment (^{1}H in both dimensions), it has a true diagonal. Cross peaks are signal intensity above the noise floor that is off of the diagonal. Cross peaks indicate coupling between the spins at the two chemical shifts of the cross peak (x and y).

The 2-D ^{1}H-^{13}C HSQC NMR spectrum is a heteronuclear NMR experiment. The HSQC spectrum has ^{1}H chemical shifts along the f_2 axis and ^{13}C shifts along the f_1 axis. A cross peak in the HSQC spectrum only occurs if a ^{1}H and a ^{13}C are strongly coupled, on the order of 140 Hz. This coupling is only found in one-bond couplings, meaning that an HSQC cross peak allows us to pair signals of ^{1}H's and ^{13}C's that are directly attached to one another. One, two, or three ^{1}H's may be bound to a ^{13}C for methine, methylene, or methyl groups, respectively. The HSQC spectrum, being derived from a heteronuclear experiment, has no true diagonal. The HSQC spectrum does, however, frequently display a smattering of signals running along a pseudo-diagonal, insofar as ^{1}H's and ^{13}C's with large chemical shifts tend to flock together. The signals of the ^{1}H and ^{13}C spins with more modest chemical shifts also tend to share HSQC cross peaks. Those HSQC cross peaks found well removed from the pseudo-diagonal are typically found in regions of a molecule where a strong electron density gradient is present. Heavy atoms such as bromine and proximate aromatic ring currents can account for these off-diagonal HSQC cross peaks. The HSQC experiment involves detecting ^{1}H signal while decoupling ^{13}C's. The large spectral window of ^{13}C signals and the lower gyromagnetic ratio of ^{13}C compared with ^{1}H requires the application of large amounts of *rf* power to the sample. Sample heating can cause chemical shifts of some NMR signals to wander relative to the conditions in which the 1-D ^{1}H and 1-D ^{13}C spectra are collected. The result of this can be that the 1-D spectra used in lieu of 2-D matrix projections may not line up well with cross peaks.

The 2-D ^{1}H-^{13}C gHMBC NMR spectrum is also a heteronuclear NMR experiment. The gHMBC spectrum will, just like the HSQC spectrum, have ^{1}H chemical shifts on the f_2 axis and ^{13}C chemical shifts on the f_1 axis. The gHMBC experiment is less sensitive than the HSQC experiment, which makes intuitive sense because the interactions we measure with this method have an average magnitude of 8 Hz compared with 140 Hz for the HSQC experiment. The difference between the gHMBC and the HSQC spectra, however, is that the gHMBC spectrum contains cross peaks between the signals of ^{1}H's and ^{13}C's separated typically by two to five bonds. In some cases, gHMBC cross peaks are observed from spins separated by more than five bonds. For example, the first molecule discussed in Chapter 5 generates a gHMBC spectrum with cross peaks from ^{1}H's and ^{13}C's separated by seven bonds (!). Although the gHMBC experiment also is ^{1}H-detected like the HSQC experiment, decoupling of the ^{13}C's is not carried out during signal detection. Because ^{1}H signal is observed with the ^{1}H's coupled to nearby ^{13}C's, the gHMBC spectrum will also contain a pair of HMQC cross peaks split by the $^1J_{CH}$. Instead of collapsing the ^{1}H signal normally found in the ^{13}C satellite peaks of the 1-D ^{1}H spectrum, the gHMBC spectrum leaves the satellite peaks where they are found in the 1-D ^{1}H spectrum and so we observe two cross peaks at with a ^{13}C chemical shift at δ_C (the same as in the HSQC spectrum), but with ^{1}H chemical shifts of $\delta_H + {^1J_{CH}}/2$ and $\delta_H - {^1J_{CH}}/2$.

Using the 1-D ^{1}H NMR spectrum, the 1-D ^{13}C NMR spectrum, and the 2-D ^{1}H-^{13}C HSQC NMR spectrum, we sort the NMR signals into four categories: (1) nonsolute molecule derived signals, (2) ^{1}H signals that do not correlate with ^{13}C signals in the HSQC spectrum, (3) ^{13}C signals that do not correlate with ^{1}H signals in the HSQC spectrum, and (4) ^{1}H and ^{13}C signals that correlate in the HSQC spectrum due to one-bond spin-spin ($^1J_{CH}$) couplings. Into category (1) we place the solvent signals and those of impurities, TMS, etc. We generally do not give these signals further consideration. Into category (2), we place the signals from ^{1}H's bound to heteroatoms. These signals are those from ^{1}H's in hydroxyl, phenolic, carboxylic, amide, and amino groups, but sometimes an alkynic ^{1}H may fail to generate an HSQC cross peak with its directly attached ^{13}C due to a one-bond coupling that is much larger than the 140-Hz expectation value upon which a refocussing delay in the HSQC pulse sequence is based. Into category (3) we place nonprotonated ^{13}C signals, and into category (4) we place paired ^{1}H and ^{13}C signals for methyl groups, methylene groups, and methine groups. For methylene groups we often have two ^{1}H signals paired with one ^{13}C signal, as molecular topology often makes geminal ^{1}H's on methylene groups occupy environments distinct from one another. Methyl and methine cross peaks in the HSQC spectrum always pair one ^{1}H chemical shift with one ^{13}C chemical shift. Borrowing a feature found in DEPT and APT spectra, the methyl- and methine-derived cross peaks often are of opposite sign compared with their methylene-derived counterparts. Depiction of processed HSQC spectra using color allows one to distinguish between positive and negative cross peaks. One can also plot contours of one sign with a small multiplicative spacing (e.g., 105%) and perhaps five contour lines, while using a more conventional 20 contours with 150% spacing for the other signed contours.

Following our sorting of signals, we examine well-resolved ^{1}H multiplets. If we know the molecular structure and can intuit the preferred molecular conformation, we can predict the expected multiplicities of the ^{1}H resonances and identify unique multiplets thereby generating additional entry points. Even narrowing a resonance down to one of two or three sites is useful. For example, two ^{1}H signals in a complex molecule may be predicted to be triplets of doublets. We can write "a or b" (in pencil, we all make mistakes) over the resonance on a copy of the ^{1}H spectrum, saving that information for later when 2-D correlations involving the resonance are considered.

We use through-bond correlation information from the COSY and the gHMBC spectrum to make additional assignments, starting with an entry point and propagating to adjacent spin sites using our understanding of bonding. In this manner, we are able to start with one or more entry points and deduce the chemical shifts of spins that couple to the spins of our known entry points.

The utility of the COSY spectrum can be negatively impacted by extensive resonance overlap. Well-resolved signals are most useful for the initial assignment of protonated species near ^{1}H signal entry points. Once tentative assignments are made, COSY cross peaks involving overlapping resonances can serve to provide weak confirmations of existing assignments. Only as a last resort do we entertain arguments that start similarly to "If we assign this resonance to site 5, then this cross peak may be from the H5-H6 coupling."

The ^{1}H's of a methyl group may couple weakly to ^{1}H's four bonds distant, especially in the case of geminal methyl groups. Recognition of $^4J_{HH}$ cross peaks in the COSY extends the range of spins we are able to assign.

We may need to measure the coupling constant between two ^{1}H resonances, both of which are overlapped with other ^{1}H signals. If this is the case we collect a phase-sensitive 2-D ^{1}H-^{1}H COSY spectrum and attempt to measure the coupling by examining f_2 slices (rows if f_1 is vertical, columns if f_1 is horizontal). The dimensions of our 2-D matrix should be adjusted so as to make the digital resolution as fine as possible in the f_2 dimension (~16 k) while simultaneously reducing the f_1 dimension (~1 k) in order to prevent the size of the matrix from becoming computationally prohibitive. The cross peaks arising from coupling that are found in a phase-sensitive 2-D COSY spectrum are modulated by the coupling responsible for the cross peak. The coupling that generates a COSY cross peak is said to be the active coupling. Other couplings in which a given spin is involved are said to be passive couplings. Measurement of a small active coupling in a resonance with larger passive couplings can be challenging. For a resonance that is a doublet, the measurement of the coupling is trivial. Measurement of a small active coupling for multiplets produced by the coupling of interest and one or more larger passive couplings can prove challenging. For example, a quartet of doublets with three large passive couplings and one small active coupling will produce slices of the data matrix along f_2 that have minima and maxima at the edges and a nearly zero middle region.

The double-quantum-filtered COSY spectrum suppresses diagonal intensity for most of the ^{1}H signals and so makes it possible to identify COSY cross peaks near the diagonal that are obscured due to their proximity to intense diagonal signals.

The signal dispersion present in the gHMBC spectrum provides a wealth of information-rich cross peaks. If a methyl group is observed to generate a ^{1}H signal that is a singlet, this generally means that the signal of the ^{1}H's in the methyl group will correlate in the gHMBC spectrum with the signals of all ^{13}C's two and three bonds distant. In additional to making use of the strong ^{1}H signals of methyl groups, we also utilize well-resolved methylene and methine ^{1}H signals. In especially rigid systems, we proceed with caution because the gHMBC spectrum may contain cross peaks resulting from $^2J_{CH}$'s, $^3J_{CH}$'s, and $^4J_{CH}$'s. Understanding of the bond alignments that promote spin-spin couplings allows us to use gHMBC cross peaks to assign the NMR signals of ^{1}H's and ^{13}C's expected to couple with known ^{13}C and ^{1}H resonances, respectively.

Corticosteroids

2.1 ANDROSTERONE IN CHLOROFORM-D

Androsterone is a natural product containing four rings, a hydroxyl group, and a carbon-oxygen double bond. Its empirical formula is $C_{19}H_{30}O_2$. The structure of androsterone is shown in Fig. 2.1.1. Table 2.1.1 shows how the methyl, methylene, methine, and nonprotonated carbon sites can be sorted based on the numbered molecular sites.

Fig. 2.1.2 shows the 1-D ^1H NMR spectrum of androsterone in chloroform-d, and Fig. 2.1.3 shows androsterone's 1-D ^{13}C NMR spectrum obtained in chloroform-d. One obvious entry point for this molecule is the nonprotonated ketone carbonyl carbon (C17), whose resonance is observed far downfield (at a high ppm value) at 221.8 ppm. Another entry point is the ^1H resonance associated with the methine at site 3, which is uniquely downfield at 4.06 ppm because its spins are just two bonds from the electronegative oxygen attached to C3.

The 1-D ^1H spectrum allows us to identify the signals of the two methyl groups at sites 18 and 19 as the strong singlets that each feature an integral corresponding to 3 hydrogens, but it does not allow us to easily determine which is which. We might speculate that site 18, being closer to oxygen than site 19, should generate a resonance more downfield (at higher ppm values), but the chemical shift differences between the signals of 18 and 19 are not profound enough to elicit a high degree of confidence in differentiating between the signals.

Fig. 2.1.4 shows the 2-D ^1H-^1H COSY NMR spectrum. Initial inspection of this spectrum shows correlations between the far downfield H3 signal and other ^1H signals, presumably from sites 2 and 4, but also possibly from the signals of sites 1 and 5.

Using the same proximity-to-oxygen argument, we can identify the methylene group at site 16 using the 2-D ^1H-^{13}C HSQC NMR spectrum (Fig. 2.1.5). Before we proceed further, we generate Table 2.1.2 to list all of ^1H and ^{13}C signals we expect to correlate through one-bond heteronuclear couplings.

Because the signal of the site 16 methylene group has a carbon shift (36.1 ppm) only 0.1 ppm downfield from another methylene ^{13}C signal (36.0 ppm), we can confirm, using the 2-D ^1H-^{13}C gHMBC spectrum, that the H16 signals share cross peaks with the C17 signal (left side of Fig. 2.1.6).

The H16 signals and the methyl ^1H resonance at 0.87 ppm share gHMBC cross peaks with the methine ^{13}C signal at 51.7 ppm, and so, by examining Fig. 2.1.1, we can reasonably infer that the 51.7 ppm shift is that of C14 signal. The ^1H methyl signal at 0.87 ppm must be from

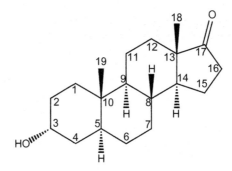

FIG. 2.1.1   The structure of androsterone.

TABLE 2.1.1   Carbons of Androsterone

Type of Carbon	Site Number
CH₃ (methyl)	18, 19
CH₂ (methylene)	1, 2, 4, 6, 7, 11, 12, 15, 16
CH (methine)	3, 5, 8, 9, 14
C_np (nonprotonated)	10, 13, 17

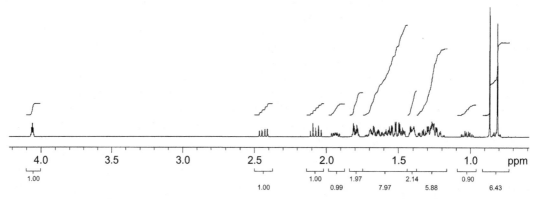

FIG. 2.1.2   The 1-D ^{1}H NMR spectrum of androsterone in chloroform-*d*.

site 18, and we can also note that these methyl ^{1}H signals share a gHMBC cross peak with the carbonyl carbon signal from site 17 (Fig. 2.1.7).

The H18 signals at 0.87 ppm share a gHMBC cross peak with a nonprotonated carbon at 48.0 ppm. This nonprotonated carbon signal must correspond to site 13, as the next nearest nonassigned, nonprotonated carbon signal is at site 10.

By the process of elimination, we can identify the signals of both the site 19 methyl group and the site 10 nonprotonated carbon. That is, we only have one of each of these types of carbons left to assign.

Using the gHMBC spectrum, we observe cross peaks between the H16 signals and a methylene ^{13}C signal at 22.0 ppm. We assign this signal to site 15.

Again using the gHMBC spectrum, we observe correlations between the two methylene signals at 1.26 & 1.80 ppm and the C13 signal at 48.0 ppm and the C18 signal at 14.0 ppm.

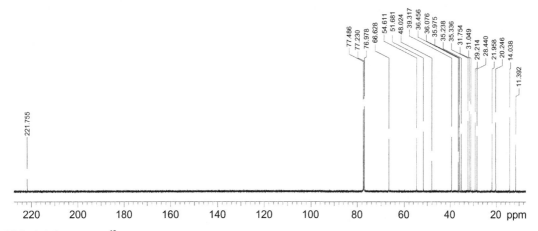

FIG. 2.1.3 The 1-D ^{13}C NMR spectrum of androsterone in chloroform-*d*.

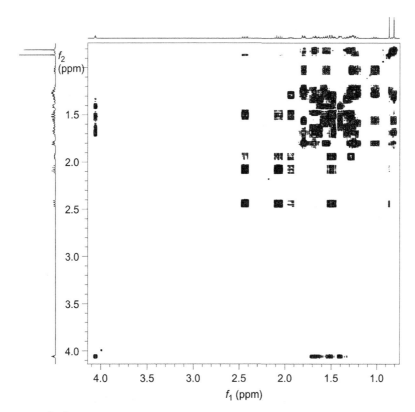

FIG. 2.1.4 The 2-D ^{1}H-^{1}H COSY NMR spectrum of androsterone in chloroform-*d*.

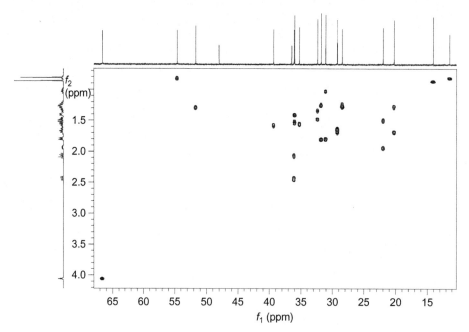

FIG. 2.1.5 The 2-D ^{1}H-^{13}C HSQC NMR spectrum of androsterone in chloroform-*d*.

TABLE 2.1.2 ^{1}H and ^{13}C NMR Signals of Androsterone Listed by Group Type

^{1}H Signal (ppm)	^{13}C Signal (ppm)	Group Type
–	221.8	Nonprotonated
4.06	66.6	Methine
0.83	54.6	Methine
1.30	51.7	Methine
–	48.0	Nonprotonated
1.58	39.3	Methine
–	36.5	Nonprotonated
2.07 & 2.43	36.1	Methylene
1.41 & 1.53	36.0	Methylene
1.56	35.2	Methine
1.34 & 1.48	32.3	Methylene
1.26 & 1.80	31.8	Methylene
1.03 & 1.80	31.0	Methylene
1.65 & 1.70	29.2	Methylene
1.24 & 1.28	28.4	Methylene
1.50 & 1.94	22.0	Methylene
1.27 & 1.69	20.2	Methylene
0.87	14.0	Methyl
0.81	11.4	Methyl

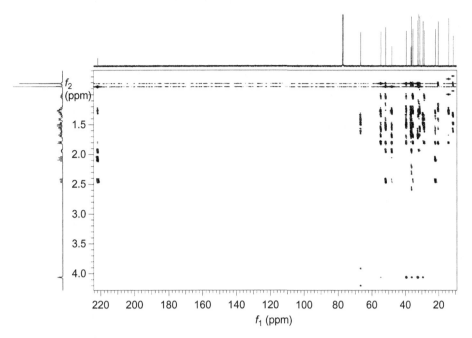

FIG. 2.1.6 The 2-D ^{1}H-^{13}C gHMBC NMR spectrum of androsterone in chloroform-*d*.

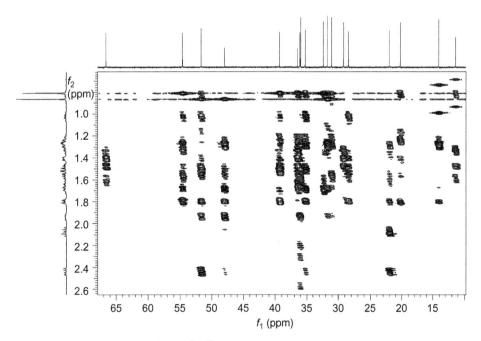

FIG. 2.1.7 An expanded portion of the 2-D ^{1}H-^{13}C gHMBC NMR spectrum of androsterone in chloroform-*d*.

We assign the 1.26 & 1.80 ppm methylene shifts to site 12. The H11 signals are unlikely to *both* share correlations with the C18 signal, because C18 is four bonds distant.

The H3 signal at 4.06 ppm shares a gHMBC cross peak with a methine carbon resonance at 39.3 ppm. Examination of the structure of androsterone reveals that the only methine group near site 3 is found at site 5 (the next nearest is at site 9, but C9 is five bonds from H3, whereas C5 is only three bonds removed from H3). The shift of the H5 signal (1.58 ppm, one of eight overlapping signals found between 1.43 ppm to 1.73 ppm) makes it difficult to use this ^{1}H shift for identifying nearby spins.

In the COSY spectrum, the well-resolved H3 signal shares correlations with two pairs of methylene resonances at 1.65 & 1.70 ppm, and at 1.41 & 1.53 ppm. H3 is not expected to share couplings with both H1's, so we can reasonably conclude that the two methylene pairs are from sites 2 and 4, but we cannot as yet determine which is which. In the gHMBC spectrum, the H3 signal of course shares a correlation with the carbon signals associated with sites 2 and 4 (already identified using the COSY spectrum and the information in Table 2.1.2 as $\delta_C = 36.0$, 29.2 ppm), but also correlates with a third methylene carbon signal at 32.3 ppm. Site 1 is the only other reasonably proximate methylene group, and so this site is interestingly established more readily than resolving the signals of sites 2 and 4 from one another. The ^{1}H signals at 1.41 & 1.53 ppm and 1.65 & 1.70 ppm from the methylene groups at sites 2 and 4 both correlate with the C3 signal at 66.6 ppm on the left side of the gHMBC spectrum in Fig. 2.1.7. Of the two pairs of methylene ^{1}H signals, the downfield pair correlate with the C1 signal at 32.3 ppm while the upfield pair correlate with the C5 signal at 39.3 ppm. We assign the 1.65 & 1.70 ppm signals to site 2 and the 1.41 & 1.53 ppm signals to site 4.

If we predict the multiplicities for the signals of H8 and H9, we determine that both of these axial ^{1}H's (we assume that each six-membered ring is in the lowest energy chair conformation) will generate signals split by multiple large (*trans* ^3J's) couplings with vicinal ^{1}H's. The H8 signal is expected to be a quartet of doublets (3 large J's, 1 small), while the H9 signal is expected to be only a triplet of doublets (there is no axial ^{1}H to which H9 can couple at site 10). While the ^{1}H methine resonance at 0.83 ppm partially overlaps with the methyl signal at 0.81 ppm (from site 19), we are still able to clearly see that it much more closely resembles a triplet of doublets than a quartet of doublets and so must arise from H9. Because site 8 is now the sole remaining methine to be identified, we assign the methine at 1.56 ppm (^{1}H) and 35.2 ppm (^{13}C) to site 8. The shift of H8 (1.56 ppm) is in a portion of the ^{1}H spectrum most subject to profound resonance overlap, and so it comes as little surprise that we arrived at this assignment through the process of elimination, rather than through some direct cross peak involving the H8 signal. Also bear in mind that the gHMBC spectrum affords a similar workaround insofar as we could have identified a gHMBC cross peak involving the C8 signal and then of course used the HSQC correlation between the signals of H8 and C8 to identify the H8 signal.

It is also interesting to note that H8 and H9 appear to be in reasonably similar environments, and yet the H9 signal is observed at 0.83 ppm while that of H8 is seen at 1.56 ppm. It is tempting to ascribe the additional shielding experienced by H8 to a pair of 1,3-diaxial interactions with the two methyl groups (sites 18 and 19), as it appears that H8 is essentially wedged between the two axial methyl groups.

Knowing the shift of the H9 signal allows us to use the well isolated resonance at 0.83 ppm to find the gHMBC cross peak shared between the signals of H9 and C11, and so we are able to identify the methylene carbon signal at 20.2 ppm as being that of site 11.

We now have two methylene groups to yet identify, those at sites 6 and 7. The extensively overlapped H8 resonance at 1.56 ppm has a characteristic width that we see repeated in the gHMBC spectrum as we move along the ^{13}C shift axis at the ^{1}H shift of 1.56 ppm. The cross peak between the signals of H8 and C9 reveals the width of the H8 resonance. We must be careful that our plotting threshold is set so as to not obscure cross peak involving ^{1}H signals with large widths. We should have a high signal-to-noise ratio spectrum and a plot with the threshold set just above the noise level so as to not miss the observation of the full width of a gHMBC cross peak along the ^{1}H axis. Equipped with the knowledge of how wide other gHMBC cross peaks should be, we proceed along the ^{13}C shift axis and encounter the same cross peak shape at 31.0 ppm.

It is possible that H8 will also couple strongly with C6, but it appears that the androsterone molecule has more conformational variation involving the 5–10 ring (ring II if numbering from lower left to upper right). If there is conformational variation for the ring on which sites 6 and 7 are found, then we expect that the ^3J's will be more adversely affected than any ^2J's, because the ^3J's depend on the varying dihedral angles while the ^2J's depend on the much more static bond angles resulting from sp^3-hybridization.

Tentative assignment of the H7 signals to 1.03 & 1.80 ppm allows us to then use these two shifts and also that of the H5 signal (1.58 ppm) to locate the C6 signal using the gHMBC spectrum. The ^{13}C resonance at 28.4 ppm shares gHMBC correlations with the signals of H5 and both putative H7's, and so we have a self-consistent and now less tentative faith that we have assigned sites 6 and 7 correctly.

The H7 signal at 1.03 ppm is also observed to correlate strongly in the gHMBC spectrum with the C14 signal at 51.7 ppm and with the C8 signal at 35.2 ppm.

2.2 HYDROCORTISONE IN PYRIDINE-D_5

Hydrocortisone (Fig. 2.2.1) is a very useful and interesting molecule. The molecule has an empirical formula of $C_{21}H_{30}O_5$ and features four rings, two carbonyls (one being conjugated), and three hydroxyl groups. As is often the case, it also contains two methyl groups occupying axial positions on its six-membered rings.

Our three preparatory tasks to complete prior to undertaking the assignment are (1) to predict ^{1}H resonance multiplicities (Table 2.2.1) based on ^2J's and ^3J's (assume hydroxyl ^{1}H's do not couple, but list them as expected ^{1}H signals anyways), (2) to make a table listing the signals we expect based on carbon type (Table 2.2.2), and (3) to make a table of observed ^{1}H and ^{13}C shifts based on the HSQC spectrum (and the ^{13}C spectrum), also listing the different types of carbons whose signals we expect to observe.

Table 2.2.1 shows the predicted multiplicities for the ^{1}H signals of hydrocortisone. We may not use more than one or two of these predicted multiplicities in carrying out the actual assignment, but having them handy allows us to preserve momentum when we are working through the assignment, whereas when one is weighing whether or not to try to use predicted multiplicities, we may elect to pursue other avenues because predicting multiplicities seems too tedious when one is within reach of completing an assignment problem (Table 2.2.3).

The spectra for hydrocortisone are shown as follows. The 1-D ^{1}H NMR spectrum is shown in Fig. 2.2.2. The 1-D ^{13}C NMR spectrum is shown in Fig. 2.2.3. The 2-D ^{1}H-^{1}H COSY NMR

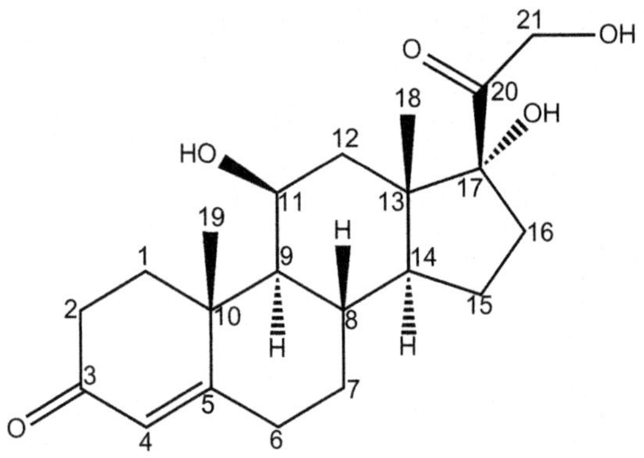

FIG. 2.2.1 The structure of hydrocortisone.

TABLE 2.2.1 Predicted Multiplicities for the ^{1}H's of Hydrocortisone

Site in Molecule	Expected Multiplicity
1	$2 \times d^3$
2	$2 \times d^3$
4	s
6	td & dt
7	qd & dq
8	qd
9	d^2
11	q
11(OH)	s
12	$2 \times d^2$
14	td
15	qd & dq
16	td & dt
17(OH)	s
18	s
19	s
21	$2 \times d$
21(OH)	s

spectrum is shown in Fig. 2.2.4. The 2-D ^{1}H-^{13}C HSQC NMR spectrum is shown in Fig. 2.2.5. The 2-D ^{1}H-^{13}C gHMBC NMR spectrum is shown in Fig. 2.2.6. An expanded portion of the gHMBC spectrum appears in Fig. 2.2.7.

The assignment will begin, as is often the case, with the entry points of ^{1}H's nearest to electronegative atoms, in this case oxygen. The C3 and C20 carbonyl signals can be easily

TABLE 2.2.2 Carbons of Hydrocortisone

Type of Carbon	Site Number
CH_3 (methyl)	18, 19
CH_2 (methylene)	1, 2, 6, 7, 12, 15, 16, 21
CH (methine)	4, 8, 9, 11, 14
C_{np} (nonprotonated)	3, 5, 10, 13, 17, 20

TABLE 2.2.3 ^1H and ^{13}C NMR Signals of Hydrocortisone Listed by Group Type

^1H Signal (ppm)	^{13}C Signal (ppm)	Group Type
–	213.6	Nonprotonated
–	199.0	Nonprotonated
–	172.6	Nonprotonated
5.85	123.0	Methine
–	90.1	Nonprotonated
4.64	68.2	Methine
4.85 & 5.31	68.1	Methylene
1.04	56.9	Methine
2.14	53.1	Methine
–	48.1	Nonprotonated
2.11 & 2.51	41.0	Methylene
–	40.1	Nonprotonated
1.83 & 2.24	35.5	Methylene
2.43 & 2.53	34.83	Methylene
1.92 & 3.19	34.77	Methylene
1.03 & 1.89	33.9	Methylene
2.11 & 2.44	32.7	Methylene
2.13	32.4	Methine
1.50 & 1.86	24.7	Methylene
1.61	21.5	Methyl
1.35	18.4	Methyl

identified from their high chemical shift values. Recalling that an isolated ketone carbonyl will generate a signal with a higher chemical shift than will a conjugated counterpart (the C4-C5 double bond makes the C3 carbonyl conjugated and able to add additional electron density through resonance), we can thus assign the most downfield shift, 213.6 ppm, to the nonconjugated C20 carbonyl signal. It then follows that C3 must generate the resonance at 199.0 ppm.

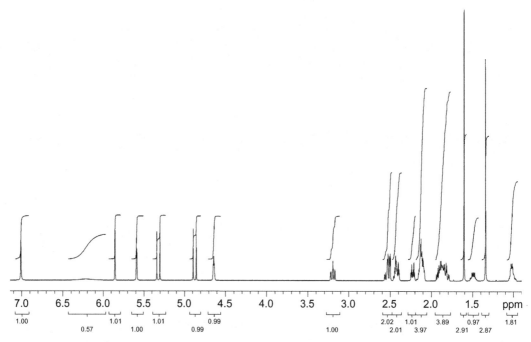

FIG. 2.2.2   The 1-D ^{1}H NMR spectrum of hydrocortisone in pyridine-d_5.

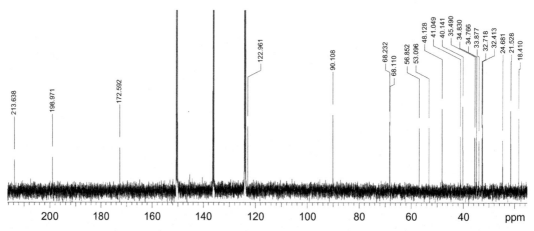

FIG. 2.2.3   The 1-D ^{13}C NMR spectrum of hydrocortisone in pyridine-d_5.

If resonance is able to add electron density to C3 and make the C3 signal shift upfield (to a lower ppm value), the same resonance withdraws electron density from C5 and so the C5 signal must be the nonprotonated carbon signal at 172.6 ppm.

The cross peak between the signals of H4 and C4 is readily apparent on the HSQC spectrum, and so we identify this easily as well. We could have started with this assignment as an alternate entry point and possibly eliminated the need to use a resonance argument to

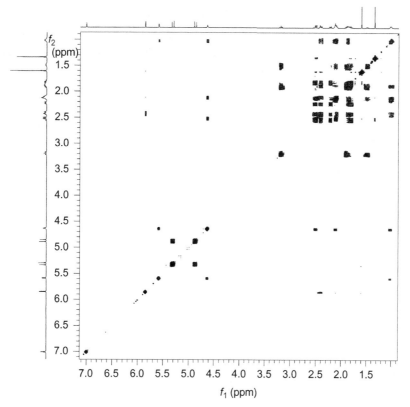

FIG. 2.2.4 The 2-D ^{1}H-^{1}H COSY NMR spectrum of hydrocortisone in pyridine-d_5.

differentiate between the signals of C3 and C20, but the notoriously small ^2J when going through the 120 degrees bond angle required by sp^2-hybridization of C4 makes the gHMBC cross peak between the signals of H4 and C3 below the plotting threshold on the gHMBC plot.

The well-resolved ^{1}H resonance of H4 does, however, share strong gHMBC cross peaks with both the signals of C2 and C10. With our knowledge of how ^3J couplings vary with dihedral angle, we predict that H4 will share strong couplings with the *trans* C2 and C10 and a moderate coupling with the *cis* C6. Examination of the gHMBC spectrum gratifyingly reveals two strong and one medium intensity cross peaks involving the ^{1}H signal at 5.85 ppm. The strong gHMBC correlation between the H4 signal (5.85 ppm) and the signal of a nonprotonated carbon at 40.1 ppm unambiguously identifies the C10 signal. The strong gHMBC correlation between the H4 signal and the methylene carbon signal at 34.8 ppm identifies the C2 signal. The ^{1}H signals from site 2 at 2.43 & 2.55 ppm correlate with the C3 signal at 199.0 ppm so we can write for site 2: δ_H = 2.43 & 2.55 ppm and δ_C = 34.83 ppm. Also, another pair of methylene ^{1}H signals at 1.83 & 2.24 ppm correlate with the C3 signal. We assign these signals to site 1: δ_H = 1.83 & 2.24 ppm and δ_C = 35.5 ppm.

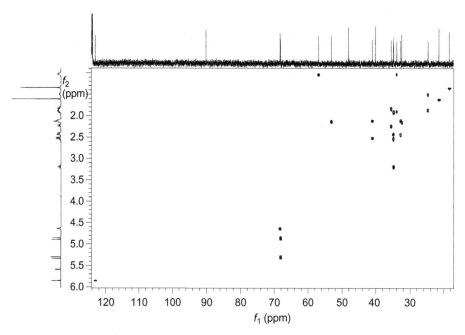

FIG. 2.2.5   The 2-D ^{1}H-^{13}C HSQC NMR spectrum of hydrocortisone in pyridine-d_5.

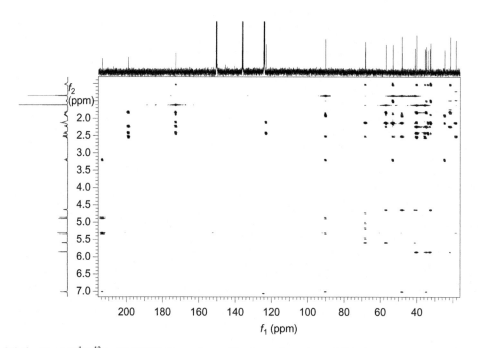

FIG. 2.2.6   The 2-D ^{1}H-^{13}C gHMBC NMR spectrum of hydrocortisone in pyridine-d_5.

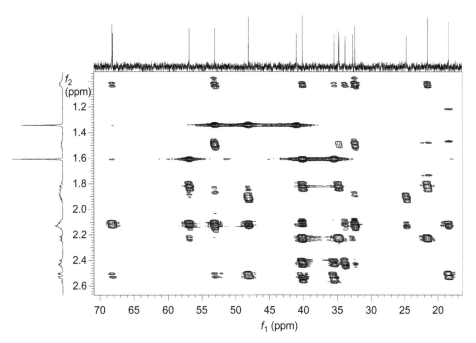

FIG. 2.2.7   An expanded portion of the 2-D ^{1}H-^{13}C gHMBC NMR spectrum of hydrocortisone in pyridine-d_5.

Finally the correlation between the H4 signal and another methylene signal at 32.7 ppm must be that attributable to the coupling between H4 and C6. If the C6 signal is found at 32.7 ppm, then the H6 signals will be those at 2.11 & 2.44 ppm. We observe gHMBC correlations between the H6 signals and that of C5 at 172.6 ppm.

Moving to the top-right portion of the molecule, we observe that a downfield methylene ^{1}H signal pair (at 4.85 & 5.31 ppm) share a strong gHMBC correlation with the C20 signal (213.6 ppm). These methylene ^{1}H signals must be from site 21, whose hydroxyl group is expected to cause its methylene ^{1}H signals to be observed at a value near 4 ppm (and with the additive effect of the carbonyl on the other side of the methylene group, it is reasonable that the H21 signals should have chemical shift values in excess of 4.0 ppm).

The COSY spectrum does not reveal a correlation between the methylene ^{1}H signals of site 21 and a hydroxyl ^{1}H signal.

The H21 signals share a second, weaker set of gHMBC correlations with a nonprotonated carbon signal at 90.1 ppm, which can only be reasonably assigned to C17 (C11 is the only other plausible candidate for that shift assignment, but C11 is protonated and so can be ruled out). Because we know C17 has a directly attached oxygen atom, we expect the C17 signal shift to be at least 60 ppm.

The methyl ^{1}H signal at 1.35 ppm shares a gHMBC cross peak with the C17 signal (90.1 ppm). The signal at 1.35 ppm must be from the H18's, three bonds from C17, as opposed to being from the H19's, which are seven bonds separated from C17.

By the process of elimination, we can now assign the H19 signals as having a chemical shift of 1.61 ppm. We can readily confirm this assignment by noting a strong gHMBC cross peak between the H19 signals (1.61 ppm) and those of C5 (172.6 ppm), because the H19's and C5 are only separated by three bonds.

Sometimes an uncoupled methyl group is situated in a way that is especially useful in the gHMBC spectrum. This is one of these cases. The ^{1}H signal found at 1.35 ppm (from site 18) participates in four gHMBC cross peaks with ^{13}C signals with shifts of 90.1 ppm (already known to be the nonprotonated C17 signal), 53.1 ppm (a methine group), 48.1 ppm (a nonprotonated carbon), and 41.0 ppm (a methylene group). The methyl group at site 18 is attached to the nonprotonated C13 and C13 is in turn attached to C12, a methylene, C14, a methine, and C17, a nonprotonated carbon we have already assigned. We expect the H18's to couple well to C13 two bonds away. Given that it is eminently reasonable (methyl groups rotate rapidly on the coupling timescale) to assume the H18's will couple equally to all three carbons that are three bonds distant, we can assign the 53.1 ppm methine signal to C14, the 48.1 ppm nonprotonated carbon signal to C13, and the 41.0 ppm methylene signal to C12. These last three carbons are all bound to C13, the nonprotonated carbon to which C18 is attached, and are all therefore three bonds distant from the H18's.

The methine ^{1}H signal at 4.64 ppm shares its strongest gHMBC cross peak with the C13 signal at 48.1 ppm. The high value of the ^{1}H chemical shift suggests that this methine is bonded to an oxygen. Only site 11 satisfies this criterion. In fact, we could have used site 11 as our initial entry point, being able to readily differentiate between the site 11 methine signal and the site 21 methylene signal using the HSQC spectrum.

The hydroxyl group on C11 has a ^{1}H signal (5.59 ppm) that shares a COSY cross peak with the H11 signal (4.64 ppm). The coupling that generates a cross peak involving the signal of a hydroxyl ^{1}H is only observed in aprotic solvents (our solvent for this molecule is pyridine-d_5). In cases where exchange occurs slowly, the hydroxyl-signal-to-methine/methylene-signal peak will only be observed directly following sample preparation and will fade as the labile protons exchange away for solvent deuterons over time.

The methylene ^{1}H signals at 1.92 & 3.19 ppm (note the large shift difference, indicating a large gradient in the chemical shielding environment) share gHMBC cross peaks with the C20 signal at 213.6 ppm, with the C17 signal at 90.1 ppm, with the C14 signal at 53.1 ppm, and with that of a methylene group whose carbon resonates at 24.7 ppm. These can only be the attributes of the site 16 methylene ^{1}H signals, with the most-recently mentioned methylene ^{13}C signal at 24.7 ppm being assigned to site 15.

All that remains to be assigned are methylene 7, and methines 8 and 9. In the COSY spectrum, we find that the H6 signals at 2.11 & 2.44 ppm share cross peaks with an as-of-yet unassigned methylene signal at 2.24 ppm, and so this must be the H7 methylene group. In the gHMBC spectrum, the H6 signals share cross peaks with the C7 methylene signal at 35.5 ppm.

We can differentiate between the ^{13}C signals of sites 8 and 9 by noting the gHMBC cross peak between the H19 signal at 1.61 ppm and to the methine ^{13}C signal at 56.9 ppm. We assign the 56.9 ppm signal to site 9: $\delta_H = 1.04$ ppm and $\delta_C = 56.9$ ppm. The site 8 methine is therefore the last unassigned group and we write for site 8: $\delta_H = 2.13$ ppm and $\delta_C = 32.4$ ppm. Unfortunately, the H9 signal is 0.01 ppm downfield from the upfield H1 signal, and the downfield H1 signal is the third of the four overlapping ^{1}H resonances at 1.83, 1.86, 1.89, and 1.92 ppm. The gHMBC cross peaks that prove to be the most informative are between the downfield H1 signal (1.89 ppm) and the C9 signal (56.9 ppm) and also the C2 signal (32.7 ppm). With hindsight, this assignment appears to be reasonable and fully consistent with our other assignments, but the high density of ^{1}H resonances would prevent us from being confident in this particular signal-to-site assignment had we made it earlier on in working this problem.

One of the H15 signals is well resolved at 1.50 ppm, and so we can use this resonance to confirm the shift of the C8 signal by noting the cross peak in the gHMBC spectrum between the upfield H15 signal (1.50 ppm) and the ^{13}C signal at 32.4 ppm (C8).

2.3 PREDNISOLONE IN N,N-DIMETHYLFORMAMIDE-D₇

A significant problem associated with a solvent such as N,N-dimethylformamide-d_7 is that having two methyl carbon signals from the solvent, where each additional deuteron splits the methyl signals into three lines, can result in a large number of ^{13}C signals that contribute nothing toward making our assignments, and so we may be tempted to completely ignore this region of the ^{13}C chemical shift axis. Sometimes a solute ^{13}C resonance can be found hidden within this region, and so we must take great care to ensure we are not missing some of our signals in the two portions of the ^{13}C chemical shift axis featuring the prominent solvent methyl signals. In this problem, there lies a ^{13}C resonance at 34.7 ppm, well inside one of the seven-line multiplets from the solvent's methyl ^{1}H signals whose intensities are in the ratio of 1:3:6:8:6:3:1.

The structure of prednisolone is shown in Fig. 2.3.1. The 1-D ^{1}H NMR spectrum of prednisolone dissolved in N,N-dimethylformamide-d_7 appears in Fig. 2.3.2. Figs. 2.3.3 and 2.3.4 show expanded portions of the 1-D ^{1}H NMR spectrum of prednisolone. The 1-D ^{13}C NMR spectrum of prednisolone appears in Fig. 2.3.5. The 2-D ^{1}H-^{1}H COSY NMR spectrum of prednisolone is found in Fig. 2.3.6. Fig. 2.3.7 contains the 2-D ^{1}H-^{13}C HSQC NMR spectrum of prednisolone, while Fig. 2.3.8 contains a portion of the spectrum that has been expanded to show detail more clearly. The 2-D ^{1}H-^{13}C gHMBC NMR spectrum of prednisolone is shown in Fig. 2.3.9. An expanded portion of the gHMBC spectrum appears in Fig. 2.3.10.

As is our custom, we generate a table showing our expected ^{1}H signal multiplicities based on ^2J's and ^3J's. Table 2.3.1 show these predictions. Table 2.3.2 shows what types of carbon-proton diads we have to assign based on their site number. Table 2.3.3 shows the information we can tabulate based on the 1-D ^{13}C NMR spectrum and the 2-D ^{1}H-^{13}C HSQC spectrum. Note that the HSQC spectrum allows us to find our obscured methylene group whose ^{13}C shift is about 34.7 ppm (our 1-D ^{13}C peak picking algorithm fails to find this signal directly).

We have two carbonyl groups in prednisolone, one of which is conjugated while the other is not. We can use our knowledge that resonance will lower the observed shift of a spin's signal by allowing the remote double bonds to donate electron density to the electronegative (due to oxygen) and electron-withdrawing (due to ability to formulate a resonance structure with a negative formal charge) carbonyl group.

The ^{13}C resonance at 212.8 ppm is attributed to the unconjugated ketone carbonyl C20, while the ^{13}C resonance at 186.4 ppm is attributed to the conjugated ketone carbonyl C3.

The resonance structures that can be drawn wherein a negative formal charge is placed on the site 3 oxygen atom require a concomitant positive charge at either site 1 (a methine group) or at site 5 (a nonprotonated carbon). Even without the HSQC spectrum, we can differentiate

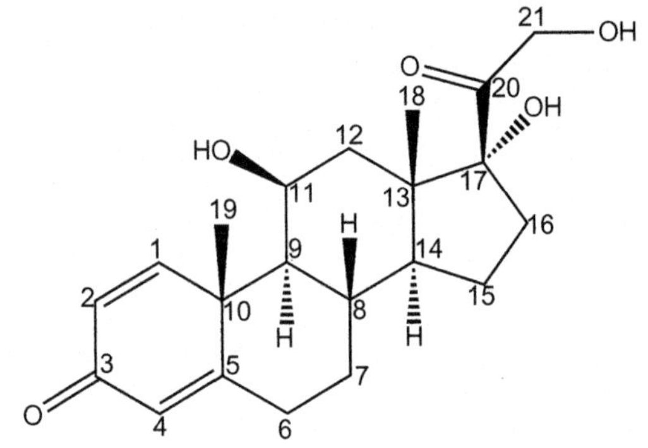

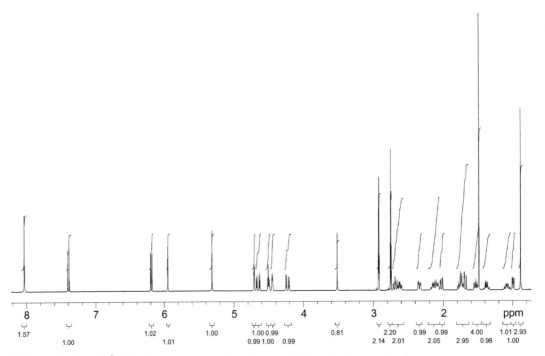

FIG. 2.3.1   The structure of prednisolone.

FIG. 2.3.2   The 1-D ^{1}H NMR spectrum of prednisolone in N,N-dimethylformamide-d_7.

between the C1 and C5 carbon signals in the 1-D ^{13}C spectrum. The protonated carbon (C1) signal at 157.7 ppm is more intense because C1 relaxes more completely in between scans and also receives an NOE enhancement (through dipolar relaxation) from H1, while the nonprotonated C5 is observed to generate a lower intensity signal at 171.6 ppm. Using the HSQC spectrum, we observe that the ^{13}C shift of C1 correlates with a ^{1}H shift of 7.38 ppm which must be that of H1.

Knowing the ^{1}H chemical shift of H1, we can use the COSY spectrum to find the H2 signal. The H1 (7.38 ppm) signal shares a COSY cross peak with the H2 signal at 6.18 ppm, and so we

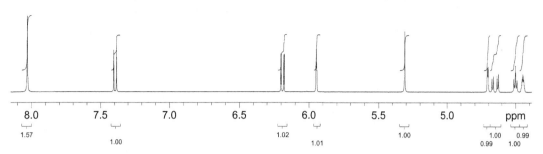

FIG. 2.3.3 The downfield portion of the 1-D ^{1}H NMR spectrum of prednisolone in N,N-dimethylformamide-d_7.

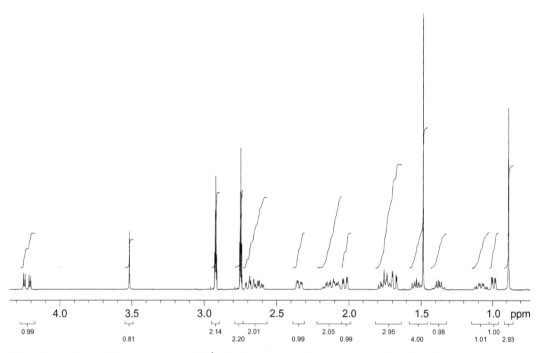

FIG. 2.3.4 The upfield portion of the 1-D ^{1}H NMR spectrum of prednisolone in N,N-dimethylformamide-d_7.

have site 2 assigned (we consult the HSQC spectrum to see the correlation between the H2 signal at 6.18 ppm and the C2 signal at 128.4 ppm).

The only other downfield methine signal is that resulting from site 4. We can identify this signal at $\delta_H = 5.94$ ppm and $\delta_C = 122.9$ ppm because it is the only remaining downfield-and-protonated signal in the HSQC spectrum (there are only three cross peaks in the HSQC spectrum with ^{13}C shifts greater than 75 ppm).

In the COSY spectrum, we observe that the H4 signal at 5.94 ppm correlates with one of the ^{1}H signals of a methylene group at 2.62 ppm, and so we assign the 2.62 ppm signal to one of the H6 sites. The other site 6 signals are $\delta_H = 2.33$ ppm and $\delta_C = 32.7$ ppm which we now can simply look up in Table 2.3.3.

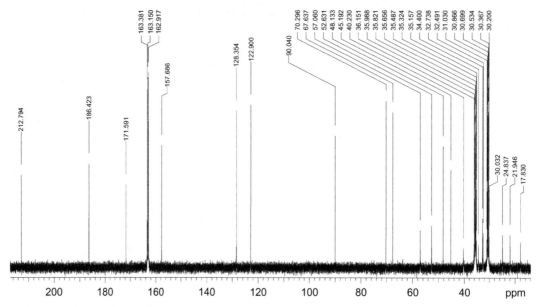

FIG. 2.3.5 The 1-D ^{13}C NMR spectrum of prednisolone in *N,N*-dimethylformamide-*d₇*.

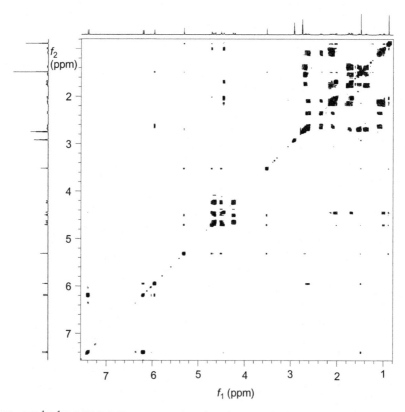

FIG. 2.3.6 The 2-D ^{1}H-^{1}H COSY NMR spectrum of prednisolone in *N,N*-dimethylformamide-*d₇*.

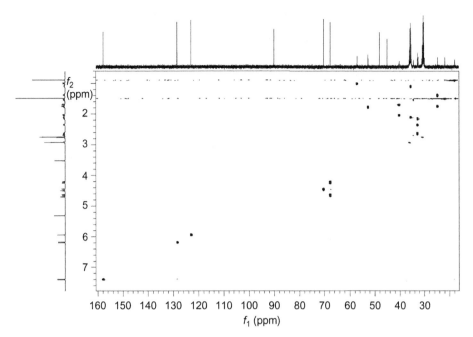

FIG. 2.3.7   The 2-D ^{1}H-^{13}C HSQC NMR spectrum of prednisolone in *N,N*-dimethylformamide-*d*$_7$.

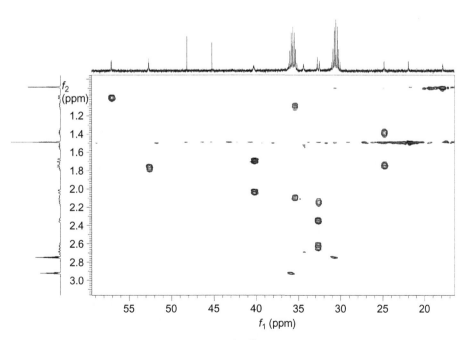

FIG. 2.3.8 An expanded portion of the 2-D ^{1}H-^{13}C HSQC NMR spectrum of prednisolone in *N,N*-dimethylformamide-*d*$_7$.

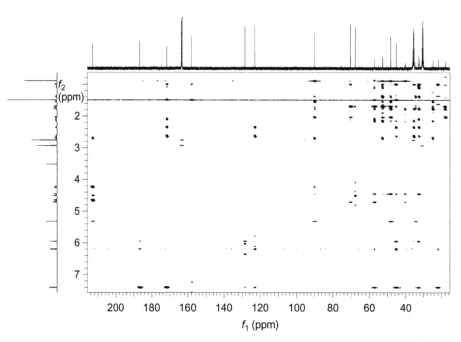

FIG. 2.3.9   The 2-D ^{1}H-^{13}C gHMBC NMR spectrum of prednisolone in *N,N*-dimethylformamide-d_7.

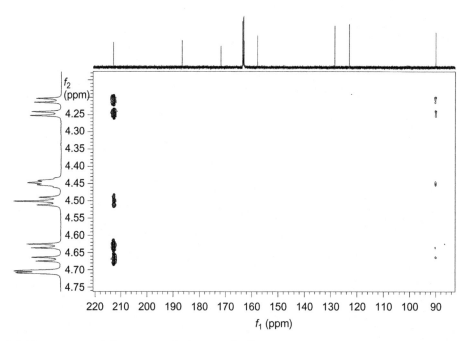

FIG. 2.3.10   An expanded portion of the 2-D ^{1}H-^{13}C gHMBC NMR spectrum of prednisolone in *N,N*-dimethylformamide-d_7.

TABLE 2.3.1   Predicted Multiplicities for the ^{1}H's of Prednisolone

Site in Molecule	Expected Multiplicity
1	d
2	d
4	s
6	td, dt
7	qd, dq
8	td
9	d^2
11	q
11(OH)	s
12	$2 \times d^2$
14	td
15	qd, dq
16	td, dt
17(OH)	s
18	s
19	s
21	$2 \times d$
21(OH)	s

TABLE 2.3.2   Carbons of Prednisolone

Type of Carbon	Site Number
CH_3 (methyl)	18, 19
CH_2 (methylene)	6, 7, 12, 15, 16, 21
CH (methine)	1, 2, 4, 8, 9, 11, 14
C_{np} (nonprotonated)	3, 5, 10, 13, 17, 20

The COSY spectrum then allows us to use the two H6 signals (2.33 & 2.62 ppm) to find cross peaks with both H7 signals. The H7 signals are found at 1.09 & 2.09 ppm, and the C7 signal is observed at 35.4 ppm.

The methyl ^{1}H signal at 1.49 ppm share a gHMBC cross peak with the C5 signal (171.6 ppm) and the C1 signal (157.7 ppm), and so the 1.49 ppm signal must be from the methyl group at site 19. The C19 signal is observed at 21.9 ppm.

TABLE 2.3.3 ^{1}H and ^{13}C NMR Signals of Prednisolone Listed by Group Type

^{1}H Signal (ppm)	^{13}C Signal (ppm)	Group Type
–	212.8	Nonprotonated
–	186.4	Nonprotonated
–	171.6	Nonprotonated
7.38	157.7	Methine
6.18	128.4	Methine
5.95	122.9	methine
–	90.0	Nonprotonated
4.45	70.3	Methine
4.21, 4.64	67.6	Methylene
1.00	57.1	Methine
1.78	52.6	Methine
–	48.1	Nonprotonated
–	45.2	Nonprotonated
1.69, 2.03	40.2	Methylene
1.09, 2.09	35.4^a	Methylene
1.53, 2.68	34.4	Methylene
2.33, 2.62	32.7	Methylene
2.14	32.5	Methine
1.38, 1.75	24.8	Methylene
1.49	21.9	Methyl
0.89	17.8	Methyl

^{a}The ^{13}C signal is overlapped with the solvent.

The H19 signals also share cross peaks in the gHMBC spectrum with ^{13}C signals at 57.1 ppm (a methine) and 45.2 ppm (a nonprotonated carbon). These two carbon signals must correspond to sites 9 and 10, respectively (note in the 1-D ^{13}C spectrum that the non-protonated signal of C10 is counterintuitively more intense than that of the site 9 methine carbon).

By the process of elimination, having identified site 19, our other methyl signal at $\delta_H = 0.89$ ppm and $\delta_C = 17.8$ ppm can be attributed to the methyl group at site 18. The H18 signal (0.89 ppm) shares gHMBC cross peaks with the ^{13}C signals at 90.4 ppm (a nonprotonated carbon), 52.6 ppm (a methine), 48.1 ppm (a nonprotonated carbon), and 40.2 ppm (a methylene). Examination of the structure of prednisolone allows us to reasonably conject that the H18's should couple to the carbons at sites 13 and 17, and we expect, based on the attachment of an oxygen atom to C17, that the shift of the C17 signal will be the larger of the two shifts of the ^{13}C signals involved in gHMBC cross peaks with those of the H18 signals. A simple rule of thumb is that a ^{13}C with a single bond to oxygen should generate a signal that has a chemical shift of at least 60 ppm (a notable exception is for the epoxide ring) and the ^{1}H or ^{1}H's attached to it should generate ^{1}H signals that have shifts near 4 ppm. This allows us to assign the ^{13}C signal at 90.0 ppm to C17 and the ^{13}C signal at 48.1 ppm to C13. The methine ^{13}C signal at 52.6 ppm we assign to C14, and the methylene ^{13}C signal at 40.2 ppm we assign to site 12.

Having identified the signal of one of the three carbons single-bonded to oxygen (C17, 90.0 ppm), we can now easily differentiate between the signals of the site 21 methylene group and the site 11 methine group (we could have used site 21 as an entry point, as it is a uniquely downfield methylene group). The site 21 methylene group's signals can be readily spotted in the middle of the HSQC figures with $\delta_H = 4.21$ & 4.64 ppm and $\delta_C = 67.6$ ppm. The site 11 methine group must therefore generate the last ^{13}C signal found with a shift value greater than 60 ppm, namely, at $\delta_H = 4.50$ ppm and $\delta_C = 70.3$ ppm.

The H11 signal (4.50 ppm) shares a gHMBC cross peak with an as-of-yet unassigned methine ^{13}C signal at 32.5 ppm, which we can assign to site 8, since we have already assigned the other nearby methines at sites 9 and 14.

All that remains to be assigned are at the two methylene groups at sites 15 and 16. The H14 signal (1.78 ppm) shares a gHMBC cross peak with the ^{13}C signal at 24.8 ppm and so we have identified the signal of C15. We verify that the C15 signal is observed at 24.8 ppm by noting that the H8 signal at 2.14 ppm shares a gHMBC cross peak with the C15 signal at 24.8 ppm, but not with the remaining ^{13}C methylene signal at 34.4 ppm, which we assign to site 16.

2.4 β-ESTRADIOL IN ACETONE-D_6

β-Estradiol is a corticosteroid whose ring I (numbering from lower left to upper right) is aromatic. The ring current generated by satisfying the $(4n + 2)\pi$ electron rule generates significant signal dispersion, and so this problem is easier to do than many of its more exclusively aliphatic corticosteroidal cousins.

The structure of β-estradiol appears in Fig. 2.4.1. The 1-D ^{1}H NMR spectrum of β-estradiol dissolved in acetone-d_6 is shown in Fig. 2.4.2. Fig. 2.4.3 contains expanded portions of the 1-D ^{1}H

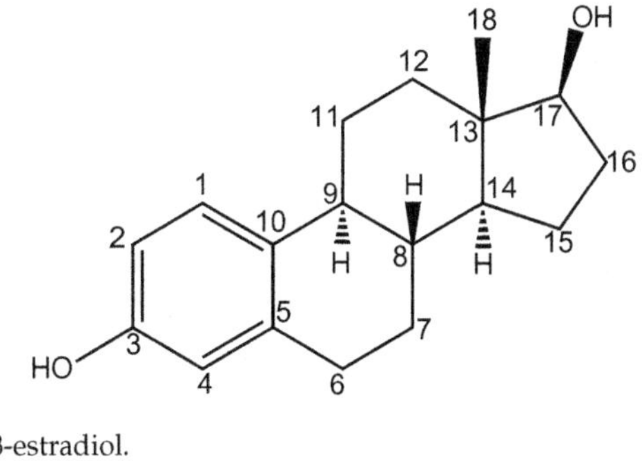

FIG. 2.4.1 The structure of β-estradiol.

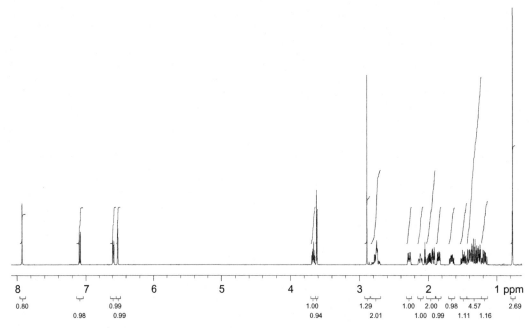

FIG. 2.4.2   The 1-D ^{1}H NMR spectrum of β-estradiol in acetone-d_6.

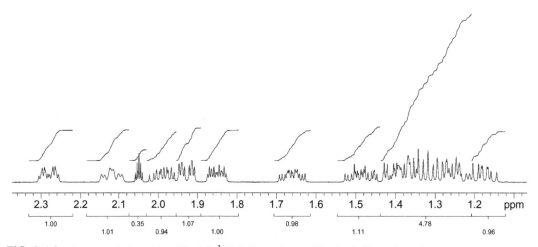

FIG. 2.4.3   An expanded portion of the 1-D ^{1}H NMR spectrum of β-estradiol in acetone-d_6.

NMR spectrum of β-estradiol. Fig. 2.4.4 contains the 1-D ^{13}C NMR spectrum of β-estradiol and Fig. 2.4.5 contains the expanded upfield portion of the same spectrum. The 2-D ^{1}H-^{1}H COSY NMR spectrum of β-estradiol is found in Fig. 2.4.6. Fig. 2.4.7 contains the 2-D ^{1}H-^{13}C HSQC NMR spectrum of β-estradiol, while Fig. 2.4.8 contains a portion of the spectrum that has been expanded to show detail more clearly. The 2-D ^{1}H-^{13}C gHMBC NMR spectrum of β-estradiol is shown in Fig. 2.4.9.

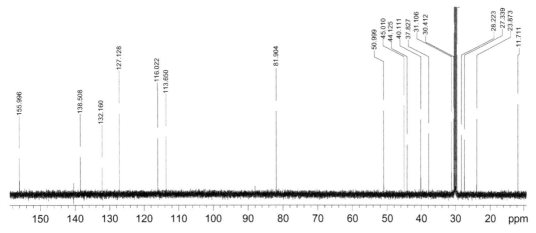

FIG. 2.4.4 The 1-D ^{13}C NMR spectrum of β-estradiol in acetone-d_6.

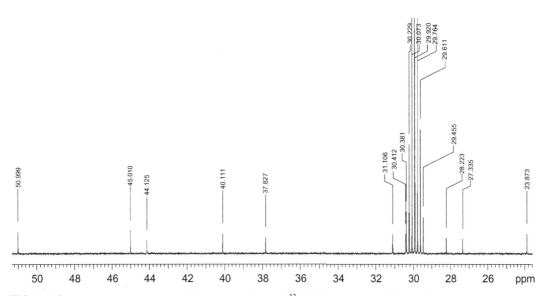

FIG. 2.4.5 An expansion of the upfield portion of the 1-D ^{13}C NMR spectrum of β-estradiol in acetone-d_6.

Initially we examine the structure of the molecule and generate our perfunctory three tables with multiplicities expected for ^{1}H resonances (Table 2.4.1), how the numbered carbon sites are partitioned based on the number of attached protons (Table 2.4.2), and the various ^{13}C signals with and without HSQC-correlated ^{1}H signals (Table 2.4.3).

Our initial entry point will be at the 3 position, because this site enjoys the unique attribute of being in the aromatic ring and having an attached oxygen atom. The ^{13}C signal from an oxygenated aromatic carbon site is typically on the order of 150 ppm—here we observe the C3 signal at 156.0 ppm.

Using our knowledge of how the lone pairs on the C3 oxygen will donate electron density through resonance to ortho ring sites 2, 4 (and para ring site 10, we expect, but other factors

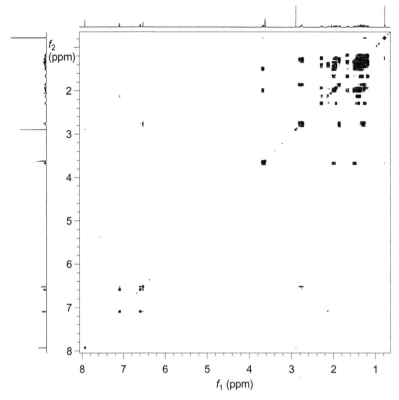

FIG. 2.4.6 The 2-D ^{1}H-^{1}H COSY NMR spectrum of β-estradiol in acetone-d_6.

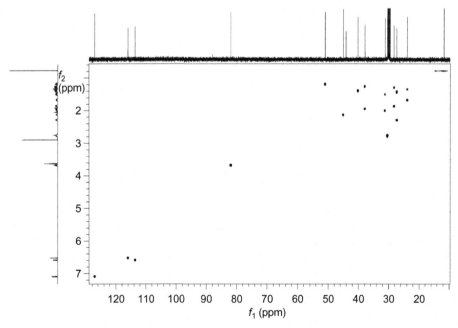

FIG. 2.4.7 The 2-D ^{1}H-^{13}C HSQC NMR spectrum of β-estradiol in acetone-d_6.

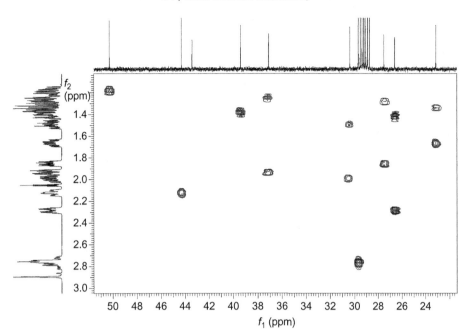

FIG. 2.4.8 An expanded portion of the 2-D ^{1}H-^{13}C HSQC NMR spectrum of β-estradiol in acetone-d_6.

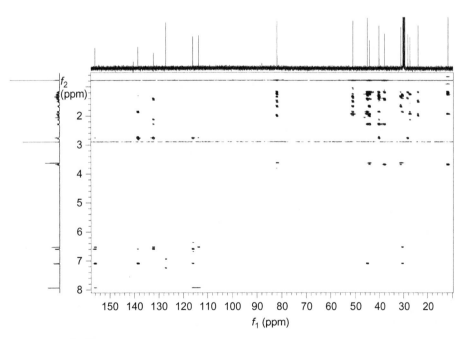

FIG. 2.4.9 The 2-D ^{1}H-^{13}C gHMBC NMR spectrum of β-estradiol in acetone-d_6.

TABLE 2.4.1   Predicted Multiplicities for the ^{1}H's of **β**-Estradiol

Site in Molecule	Expected Multiplicity
1	d
2	d
4	s
6	td, dt
7	qd, dq
8	qd
9	td
11	qd, dq
12	td, dt
14	td
15	qd, dq
16	qd, dq^a
17	d^2
18	s

^{a}*Actual multiplet likely to not fully resemble predicted multiplicity because of ring being five-membered.*

TABLE 2.4.2   Carbons of **β**-Estradiol

Type of Carbon	Site Number
CH_3 (methyl)	18
CH_2 (methylene)	6, 7, 11, 12, 15, 16
CH (methine)	1, 2, 4, 8, 9, 14, 17
C_{np} (nonprotonated)	3, 5, 10, 13

are involved), we can predict that the signals of C2 and C4 will be found at a lower ppm value than those of their meta counterparts, C1 and C5.

We can easily identify the H4 signal because it minimally split owing to H4 being four or more bonds distant from other ^{1}H's. The H4 signal is found at 6.52 ppm. The cross peak between the signals of the H1/H2 pair is easy to identify on the COSY spectrum (and with a practiced eye—on the 1-D ^{1}H spectrum—look for the intensity skewing due to the roof/Dach effect). H2 shares a weaker ^4J with H4 and so the height of the H2 resonance in the 1-D ^{1}H spectrum is lower than for the H1 signal, yet its integrated intensity is nearly the same. Thus H2 must generate the ^{1}H signal at 6.59 ppm and H1 must generate the signal at 7.09 ppm. Note that the H2 signal has a ^{1}H chemical shift on the low end of what we expect from an

TABLE 2.4.3 ^{1}H and ^{13}C NMR Signals of **β**-Estradiol Listed by Group Type

^{1}H Signal (ppm)	^{13}C Signal (ppm)	Group Type
–	156.0	Nonprotonated
–	138.5	Nonprotonated
–	132.2	Nonprotonated
7.09	127.1	Methine
6.52	116.0	Methine
6.59	113.6	Methine
3.67	81.9	Methine
1.19	51.0	Methine
2.13	45.0	Methine
–	44.1	Nonprotonated
1.41	40.1	Methine
1.24, 1.96	37.8	Methylene
1.50, 2.07	31.1	Methylene
2.76, 2.78	30.4	Methylene
1.28, 1.87	28.2	Methylene
1.43, 2.28	27.3	Methylene
1.35, 1.67	23.9	Methylene
0.77	11.7	Methyl

aromatic ^{1}H. From the HSQC spectrum we readily obtain the ^{13}C shifts for the signals of sites 1, 2, and 4 (see Table 2.4.3).

We know with the planar aromatic ring and a favorable geometry for a number of *trans*-^3J couplings, we expect to observe strong gHMBC cross peaks between the signals of H2 and C10, between those of H4 and C10, between those of H1 and C3, and between those of H1 and C5. We first use the H1 signal at 7.09 ppm to confirm our C3 signal assignment (156.0 ppm) in the gHMBC spectrum, and then are able to immediately identify the C5 signal at 138.5 ppm because we know this ^{13}C shift corresponds to a nonprotonated carbon site. Finally, we see that the H2 (6.59 ppm) and H4 (6.52 ppm) signals both share a strong gHMBC cross peak with the nonprotonated ^{13}C signal at 132.2 ppm, which must be that of C10.

We can wring some additional information from our aromatic ring. H1 and H4 are both expected to couple via a moderately strong *cis*-^3J to C9 and C6, respectively. If we examine the lower-right portion of the gHMBC spectrum, we observe three cross peaks involving two ^{13}C signals at 45.0 and 30.4 ppm. The 45.0 ppm signal is from a nonprotonated carbon site and so must be that of C9. The 30.4 ppm signal is from a methylene group and so much be that of C6. The cross peak between the signals of H1 and C4 arises from a ^4J that occurs in especially rigid and planar systems, and in this geometry is sometimes called a "W" coupling, but some purists are reluctant to label an aromatic ^4J as a W-coupling, instead reserving the term for saturated regions of molecules only.

Having exhausted all signals from spins in the vicinity of the aromatic ring, we move on and note that C17 is the only carbon besides C3 bearing an oxygen atom. In the very center of the HSQC figure we observe the cross peak between the signals of H17 and C17 with $\delta_H = 3.67$ ppm and $\delta_C = 81.9$ ppm.

A similarly unique molecular feature of β-estradiol is that it has but one methyl group, and so the methyl signal at site 18 is noted as having $\delta_H = 0.77$ ppm and $\delta_C = 11.7$ ppm.

Having already identified two of the three nonprotonated carbons (C5 and C10, both aromatic), we can use the information in Table 2.4.3 to determine that the last remaining nonprotonated ^{13}C signal (site 13) can be found at 44.1 ppm.

We have two methine groups left to identify from the spins at sites 8 and 14. The ^1H signals from the methylene group at site 6 (2.76 & 2.78 ppm) share gHMBC cross peaks with only two aliphatic ^{13}C signals at 40.1 (a methine) and 28.2 ppm (a methylene). These two carbon signals must be from sites 8 and 7, respectively. Note that the lack of planarity in this portion of the molecule means that the methine group whose ^{13}C signal appears at 40.1 ppm is highly unlikely to be from C14 (we already know the shift of the signal from C9).

By elimination we know that for the methine at site 14, $\delta_H = 1.19$ ppm and $\delta_C = 51.0$ ppm.

We now have four methylene groups to still identify at sites 11, 12, 15, and 16. The well-resolved H14 signal at 1.19 ppm shares gHMBC cross peaks with carbon signals at 45.0 ppm (C9), 44.1 ppm (C13), 40.1 ppm (C8), 37.8 ppm (a methylene group), 28.2 ppm (C7), 23.9 ppm (a methylene), and 11.7 ppm (C18). Because H14 can conceivably couple with C12, C15, and C16, we must obtain some additional information before we can proceed with our assignments. The ^1H signal at 2.28 ppm, part of one of our as-of-yet unassigned methylene groups, shares a weak cross peak with the C10 signal in the gHMBC spectrum. This ^1H signal also shares gHMBC cross peaks with ^{13}C signals at 45.0 ppm (C9), 44.1 ppm (C13), 40.1 ppm (C8), and with 37.8 ppm (a methylene group). Only one of the H11's can reasonably be expected to couple to C10, C9, C13, and C8, and so the other methylene ^{13}C signal at 37.8 ppm must be that of site 12.

The H17 signal at 3.67 ppm shares gHMBC cross peaks with ^{13}C signals at 44.1 ppm (C13), 37.8 ppm (C12), 31.1 ppm (either C15 or C15), and 11.7 ppm (C18). Given that the fluxional nature of the five-membered ring at the upper-right portion of the molecule may make ^3J's vary over time, we attribute the 31.1 ppm methylene ^{13}C signal to site 16.

The H14 ^1H signal at 1.19 ppm is found, in the gHMBC spectrum, to correlate to the ^{13}C signals of C17 (81.9 ppm), C9 (45.0 ppm), C13 (44.1 ppm), C8 (40.1 ppm), C12 (37.8 ppm), and finally to what can only be the C15 signal at 23.9 ppm.

2.5 CHENODEOXYCHOLIC ACID IN PYRIDINE-D_5

Chenodeoxycholic Acid is a corticosteroid with four rings, two hydroxyl groups, and a five-carbon side chain that terminates in a carboxyl group coming off the five-membered ring. The empirical formula for chenodeoxycholic acid is $C_{24}H_{40}O_4$. The structure of chenodeoxycholic acid is shown in Fig. 2.5.1.

A small amount of chenodeoxycholic acid was dissolved in pyridine-d_5 and this sample was used to collect the spectra appearing in this section. The 1-D ^1H NMR spectrum of

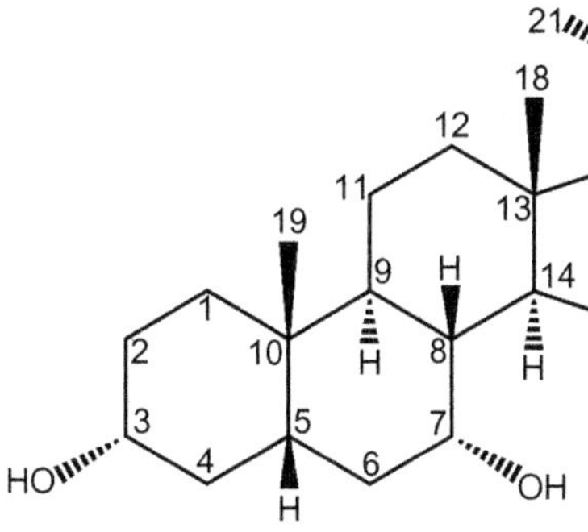

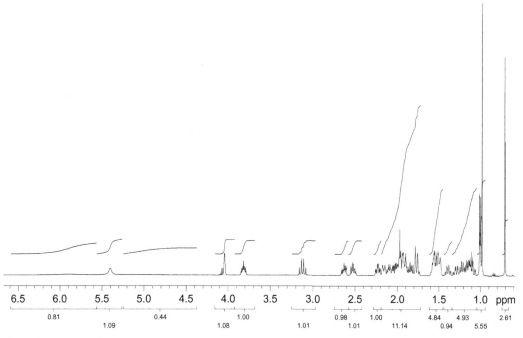

FIG. 2.5.1 The structure of chenodeoxycholic acid.

chenodeoxycholic acid in pyridine-d_5 is shown in Fig. 2.5.2. Fig. 2.5.3 contains the 1-D ^{13}C NMR spectrum of chenodeoxycholic acid. The 2-D ^{1}H-^{1}H COSY NMR spectrum of chenodeoxycholic acid is found in Fig. 2.5.4, while an expanded portion of the COSY spectrum is shown in Fig. 2.5.5. Fig. 2.5.6 contains the 2-D ^{1}H-^{13}C HSQC NMR spectrum of chenodeoxycholic acid. The 2-D ^{1}H-^{13}C gHMBC NMR spectrum of chenodeoxycholic acid is shown in Fig. 2.5.7.

We generate our expected ^{1}H multiplicity table (Table 2.5.1), our table of carbon types (Table 2.5.2), and our table listing the observed carbon signals, along with any HSQC-correlated

FIG. 2.5.2 The 1-D ^{1}H NMR spectrum of chenodeoxycholic acid in pyridine-d_5.

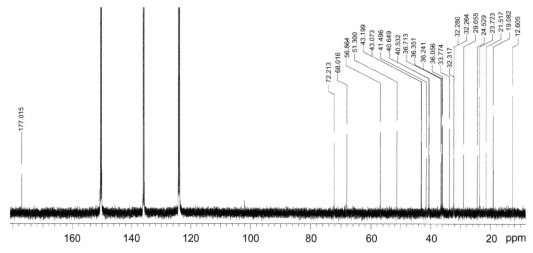

FIG. 2.5.3 The 1-D ^{13}C NMR spectrum of chenodeoxycholic acid in pyridine-d_5.

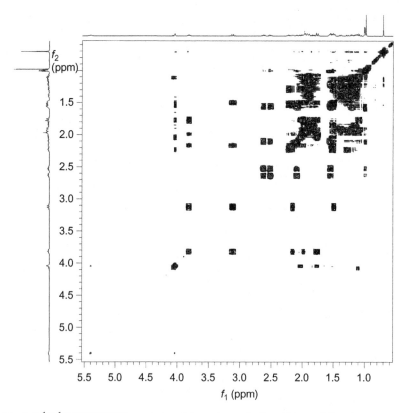

FIG. 2.5.4 The 2-D ^{1}H-^{1}H COSY NMR spectrum of chenodeoxycholic acid in pyridine-d_5.

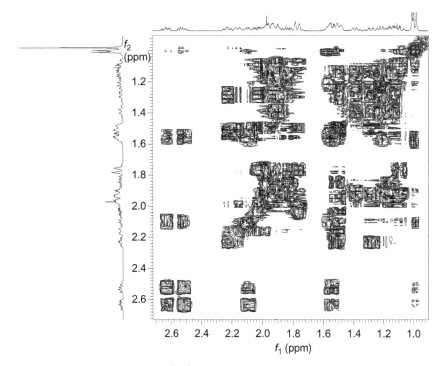

FIG. 2.5.5   An expanded portion of the 2-D ^{1}H-^{1}H COSY NMR spectrum of chenodeoxycholic acid in pyridine-d_5.

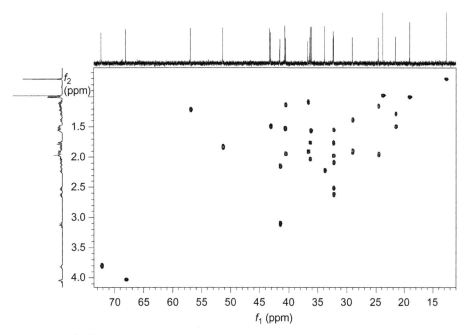

FIG. 2.5.6   The 2-D ^{1}H-^{13}C HSQC NMR spectrum of chenodeoxycholic acid in pyridine-d_5.

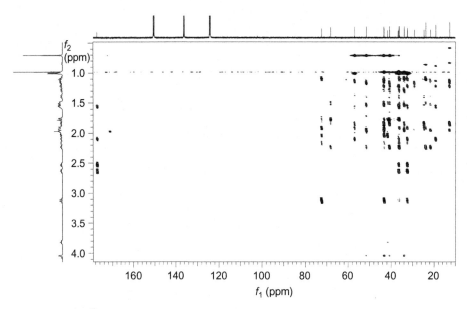

FIG. 2.5.7   The 2-D ^{1}H-^{13}C gHMBC NMR spectrum of chenodeoxycholic acid in pyridine-d_5.

^{1}H signals (Table 2.5.3). Note that multiplet prediction is complicated by the skew-boat conformation of ring I (contains carbons 1, 2, 3, 4, 5, 10), owing to the H5 and the methyl group (C19) attached to C10 being on the same face of the molecule.

With a single carbonyl carbon present, we can readily identify the ^{13}C resonance at 177.0 ppm as being that arising from C24. In the 2-D ^{1}H-^{13}C gHMBC spectrum, we observe a correlation between the C24 signal and the most downfield methylene ^{1}H signals whose shifts are 2.52 & 2.62 ppm. We assign these shifts to the site 23 methylene group and use information in Table 2.5.3 (derived from the 2-D ^{1}H-^{13}C HSQC spectrum) to determine that the C23 signal has a ^{13}C shift of 32.28 ppm. Note that the C23 signal is very close to two other signals (at 32.32 and 32.26 ppm—we have had to resort to listing these shifts to the nearest hundredth of a ppm in order to distinguish them from one another).

The ^{1}H signal of a second methylene group, which must be from site 22, also shares gHMBC correlations with the downfield C24 signal (177.0 ppm). The ^{1}H shifts of this group are 1.56 & 2.09 ppm, and from Table 2.5.3 we obtain the ^{13}C shift for this methylene group of 32.26 ppm. The close proximity of the signals of C22 and C23 along the ^{13}C chemical shift axis suggests that the gHMBC spectrum will be of little help near $\delta_C = 32.3$ ppm.

In the gHMBC spectrum, the well-resolved H23 signals (2.52 & 2.62 ppm) of course share cross peaks with the C24 signal at 177.0 ppm, but also with ^{13}C signals near ~32 ppm and ~36 ppm. Since all of the ^{13}C signals near 32 ppm arise from methylene groups (see Table 2.5.3) and since only one of the four ^{13}C signals near 36 ppm is from a methine group, we can comfortably assert that the methine ^{13}C signal at 36.1 ppm must be that from site 20, and so we consult Table 2.5.3 and determine that the shift for the H20 methine signal is 1.57 ppm.

The methyl group at site 21 should generate the only ^{1}H methyl signal that is split into a doublet (all the other methyl groups—sites 18 and 19—are bound to nonprotonated carbons).

TABLE 2.5.1 Predicted Multiplicities for the ^{1}H's of Chenodeoxycholic Acid

Site in Molecule	Expected Multiplicity
1	$2 \times d^3$
2	$2 \times d^4$
3	d^4 (OH s)
4	d^2
5	d^3
6	$2 \times d^3$
7	d^3 (OH s)
8	td
9	td
11	qd, dq
12	td, dt
14	td
15	qd, dq^a
16	qd, dq^a
17	d^3
18	s
19	s
20	qd^3
21	d
22	$2 \times d^4$
23	$2 \times d^3$

^{a}Actual multiplet may not fully resemble predicted multiplicity because of ring being five-membered.

TABLE 2.5.2 Carbons of Chenodeoxycholic Acid

Type of Carbon	Site Number
CH_3 (methyl)	18, 19, 21
CH_2 (methylene)	1, 2, 4, 6, 11, 12, 15, 16, 22, 23
CH (methine)	3, 5, 7, 8, 9, 14, 17, 20
C_{np} (nonprotonated)	10, 13, 24

The methyl doublet at 1.00 ppm is readily identified in the 1-D ^{1}H spectrum and also in the 2-D ^{1}H-^{13}C HSQC spectrum (recall that methyl cross peaks tend to be very strong and generate significant t_1 ridges in the HSQC and gHMBC spectra). The ^{13}C shift of the C21 methyl signal is 19.1 ppm. The 2-D ^{1}H-^{1}H COSY spectrum confirms our assignment of the signals of H20 and H21 by featuring a cross peak between signals at 1.57 and 1.00 ppm, although

TABLE 2.5.3 ^{1}H and ^{13}C NMR Signals of Chenodeoxycholic Acid Listed by Group Type

[1]H Signal (ppm)	[13]C Signal (ppm)	Group Type
–	177.0	Nonprotonated
3.80	72.2	Methine
4.03	68.0	Methine
1.21	56.9	Methine
1.84	51.3	Methine
–	43.2	Nonprotonated
1.49	43.1	Methine
2.16 & 3.10	41.5	Methylene
1.53	40.6	Methine
1.14 & 1.95	40.5	Methylene
1.09 & 1.91	36.7	Methylene
1.76 & 2.04	36.4	Methylene
–	36.2	Nonprotonated
1.57	36.1	Methine
2.22	33.8	Methine
1.77 & 1.98	32.32	Methylene
2.52 & 2.62	32.28	Methylene
1.56 & 2.09	32.26	Methylene
1.39 & 1.90	29.1	Methylene
1.16 & 1.96	24.5	Methylene
0.98	23.7	Methyl
1.28 & 1.50	21.5	Methylene
1.00	19.1	Methyl
0.70	12.6	Methyl

resonance overlap near 1.57 ppm is extensive, diminishing the value of this correlation. When we are confirming an assignment, we can accept a slightly higher amount of ambiguity—if we fail to accept any at all, we will find ourselves unable to assign signals in overlapped regions. Sometimes we must simply accept one set of assumptions and be prepared to entertain other options should our initial assumptions prove incorrect.

Only one other ^{1}H resonance (at 1.21 ppm) shares a gHMBC cross peak with the ^{13}C signal from the methyl group at site 21 (at 19.1 ppm). This ^{1}H signal is from a methine group with a ^{13}C shift of 56.9 ppm and these methine signals must be from site 17, as site 14 is the next nearest methine group and H14 is five bonds from C21, while H17 is only three bonds removed from C21.

Confirmation of the assignment of the site 17 signals is attained when we note that a methyl ^{1}H signal (a singlet) at 0.70 ppm shares a gHMBC cross peak with the C17 methine signal at 56.9 ppm. The methyl signal at 0.70 ppm must be that from the methyl group at site 18. The shift of the site 18 methyl ^{13}C is 12.6 ppm (obtained from the HSQC spectrum or Table 2.5.3).

In the gHMBC spectrum, we anticipate that the H18 signal will correlate with the ^{13}C signals of the known C17, as well as with the methylene signal of C12, the methine signal of C14, and the nonprotonated signal of C13. The H18 signal correlates with a methine ^{13}C signal at 51.3 ppm and so we write for site 14: δ_H = 1.84 ppm and δ_C = 51.3 ppm. The H18 signal also correlates with a nonprotonated ^{13}C signal at 43.2 ppm. We rule out the ^{13}C signal at 43.1 ppm as being that of a methine group because we have already identified the signal of the site 14 methine group. We assign the nonprotonated ^{13}C signal to site 13: δ_C = 43.2 ppm. Finally, the H18 signal correlates with a methylene ^{13}C signal at 40.5 ppm (not the methine signal at 40.6 ppm because we require a correlation to a methylene ^{13}C signal) and so we write for site 12: δ_H = 1.14 & 1.95 ppm and δ_C = 40.5 ppm.

The H14 signal at 1.84 ppm shares a gHMBC cross peak with a methylene ^{13}C signal at 24.5 ppm. Site 15 is the only methylene group adjacent to site 14, and so we write for site 15: δ_H = 1.16 & 1.96 ppm and δ_C = 24.5 ppm. One of the H15 resonances is well-resolved, and we use the ^{1}H signal at 1.16 ppm to located a gHMBC correlation with a methylene ^{13}C signal at 29.1 ppm. Because the next nearest methylene [13]C (site 12) is four bonds from the H15's, we assign this ^{13}C methylene signal to site 16: δ_H = 1.39 & 1.90 ppm and δ_C = 29.1 ppm. We obtain confirmation in the form of a gHMBC cross peak correlating the H16 signal at 1.39 ppm with the C17 signal at 56.9 ppm, thus completing our assignment of the five-membered ring of chenodeoxycholic acid.

Retracing our steps, we return to site 14. In the COSY spectrum, the H14 signal at 1.84 ppm correlates with a ^{1}H signal at 1.49 or 1.50 ppm. There are two ^{1}H signals near 1.50 ppm, one from a methylene group with a ^{13}C chemical shift of 21.5 ppm, and the other from a methine group with a ^{13}C chemical shift of 43.1 ppm. Given that the methylenes of sites 6 and 11 are both more removed from site 14 than is site 8, we assume the COSY correlation in question is that between the H14 signal at 1.84 ppm and that of the site 8 methine [1]H. We observe a confirming cross peak between the H14 signal and the ^{13}C signal of the methine group at 43.1 ppm. We write for site 8: δ_H = 1.49 ppm and δ_C = 43.1 ppm. The methine we just assigned cannot be from site 7 because C7 is alpha to oxygen. The methine NMR signals we have just assigned to site 8 could possibly be from site 9. Of the remaining unassigned methine groups (3, 5, 7, and 9), we expect both 3 and 7, because they are alpha to oxygen, to have midfield ^{13}C signal shifts. The H14 signal at 1.84 ppm also correlates in the gHMBC spectrum with a ^{13}C signal near one of the methine ^{13}C signals at 33.8 ppm, although ^{13}C signal overlap near 32–33 ppm is extensive, with three other methylene groups contributing ambiguity. The ^{1}H chemical shift of the methine with the ^{13}C shift of 33.8 ppm is 2.22 ppm, a piece of information we will use momentarily.

The putative H8 signal at 1.49 ppm is observed to share a gHMBC cross peak with a ^{13}C signal at 68.0 ppm. This midfield shift, downfield from aliphatic hydrocarbon shifts, suggests that this ^{13}C is alpha to oxygen. We assign this signal to site 7: δ_H = 4.03 ppm and δ_C = 68.0 ppm. Had we assigned the 1.49 ppm methine signal to H9, we would be at a loss to explain how H8 could generate the ^{1}H signal at 2.22 ppm which does not correlate in the gHMBC spectrum with what we take to be the C7 signal at 68.0 ppm. That is, if the H9 signal is observed at 1.49 ppm and the H8 signal at 2.22 ppm, then our observed gHMBC correlations of the signals of H9 but not H8 with the midfield C7 signal are challenging (nearly impossible) to explain. We write for site 9: δ_H = 2.22 ppm and δ_C = 33.8 ppm. The H7 signal at 4.03 ppm shares COSY cross peaks with methylene ^{1}H signals at 1.76 & 2.04 ppm. The ^{1}H methylene signals are assigned to site 6: δ_H = 1.76 & 2.04 ppm and δ_C = 36.4 ppm. No other methylene group is sufficiently close to reasonably compete for consideration. Having identified the site 7 ^{13}C shift, we can now turn our attention to site 3, which, since we have already assigned C24, should with its proximity to oxygen possess the most downfield and still unassigned ^{13}C chemical shift. Consulting Table 2.5.3, we find a midfield methine ^{1}H/^{13}C signal pair in the second row and write for site 3: δ_H = 3.80 ppm and δ_C = 72.2 ppm. The H3 signal at 3.80 ppm shares four COSY cross peaks with two pairs of methylene ^{1}H signals which we take as the H2 and H4 signals. One pair is found at 1.77 & 1.98 ppm and the other at 2.16 & 3.12 ppm. We can proceed with the simple argument that the H4 signals should have larger chemical shifts than the H2 signals. We write for site 4: δ_H = 2.16 & 3.12 ppm and δ_C = 41.5 ppm. The more upfield methylene ^{1}H signals are observed to correlate in the COSY spectrum with two other methylene ^{1}H signals at 1.09 & 1.91 ppm, indicating two methylene groups adjacent to one another. We assign these two methylenes to sites 2 and 1. For site 2 we write: δ_H = 1.77 & 1.98 ppm and δ_C = 32.32 ppm, and for site 1: δ_H = 1.09 & 1.91 ppm and δ_C = 36.7 ppm. By the process of elimination, we assign the last methyl ^{1}H/^{13}C signal pair to site 19: δ_H = 0.98 ppm and δ_C = 23.7 ppm. We are also able to uniquely identify the last nonprotonated ^{13}C signal at 36.2 ppm as being that of site 10: δ_C = 36.2 ppm. We are fortunate to be able to assign both sites 19 and 10 by the process of elimination, because overlap along the ^{13}C chemical shift axis would otherwise make this task more challenging. The last remaining methine group is at site 5: δ_H = 1.53 ppm and δ_C = 40.6 ppm. The last remaining methylene group is at site 11: δ_H = 1.28 & 1.50 ppm and δ_C = 21.5 ppm. The H9 signal at 2.22 ppm and one of the H12 signals at 1.95 ppm share strong gHMBC correlations with the C11 signal at 21.5 ppm.

2.6 ANDROGRAPHOLIDE IN PYRIDINE-D$_5$

Andrographolide may not appear to be in the same corticosteroid category as the other molecules here in this chapter because it only has three rings, but examination of its structure reveals extensive structural similarities to the other compounds in this chapter. Andrographolide has five oxygen atoms, two of which are involved in lactone functionalities (cyclic ester) with the other three oxygens appearing in hydroxyl groups. Besides the ester carbonyl group, two other double bonds are present, one of which is an exotic terminal alkene. Andrographolide also has two methyl groups and many chiral centers (3, 4, 5, 9, 10, 12/13, and 18). The empirical formula of andrographolide is $C_{20}H_{30}O_5$. The structure of andrographolide is shown in Fig. 2.6.1.

A small amount of andrographolide was dissolved in pyridine-d_5. This sample was used to collect the spectra appearing in this section. The 1-D ^{1}H NMR spectrum of andrographolide in pyridine-d_5 is found in Fig. 2.6.2. Fig. 2.6.3 shows the 1-D ^{13}C NMR spectrum of andrographolide. The 2-D ^{1}H-^{1}H COSY NMR spectrum of andrographolide appears in Fig. 2.6.4. Fig. 2.6.5 contains the 2-D ^{1}H-^{13}C HSQC NMR spectrum of andrographolide. The 2-D ^{1}H-^{13}C gHMBC NMR spectrum of andrographolide is shown in Fig. 2.6.6.

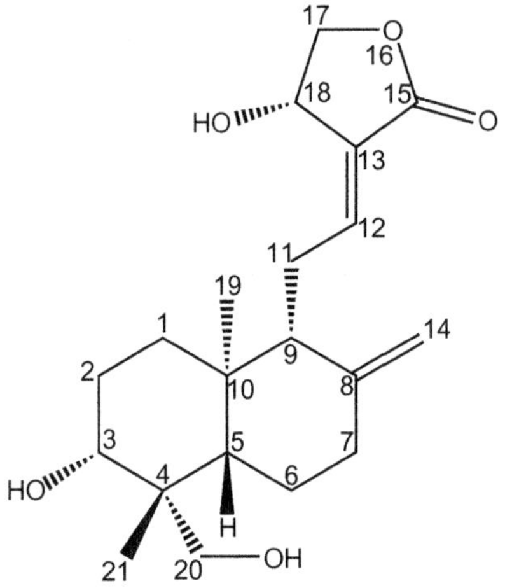

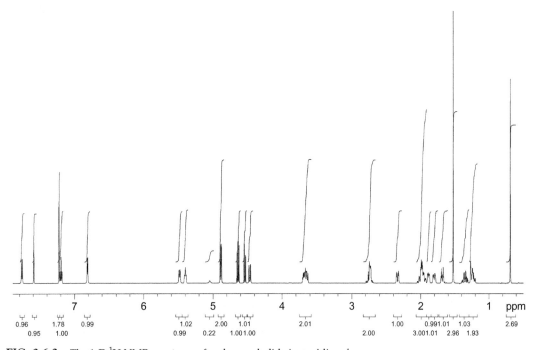

FIG. 2.6.1 The structure of andrographolide.

FIG. 2.6.2 The 1-D ^{1}H NMR spectrum of andrographolide in pyridine-d_5.

As is standard, we produce an expected ^{1}H multiplicity table (Table 2.6.1), a table of carbon types (Table 2.6.2), and a table listing the observed carbon signals, along with any HSQC-correlated ^{1}H signals (Table 2.6.3). Note that we cannot immediately know the conformation of the C9-C11, and C11-C12 bonds, and so our ability to accurately predict multiplets based on $^3J_{HH}$'s resulting from specific dihedral angles is curtailed. We can assume free rotation of the two aforementioned bonds and use intermediate couplings when we predict resonance multiplicities.

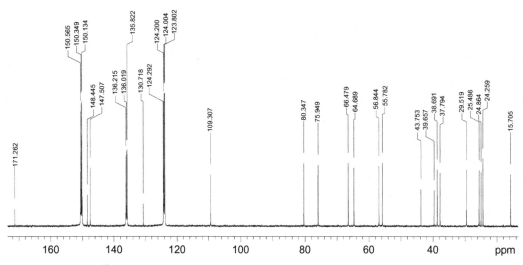

FIG. 2.6.3   The 1-D ^{13}C NMR spectrum of andrographolide in pyridine-d_5.

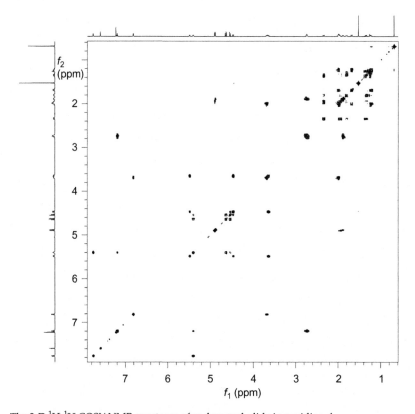

FIG. 2.6.4   The 2-D ^{1}H-^{1}H COSY NMR spectrum of andrographolide in pyridine-d_5.

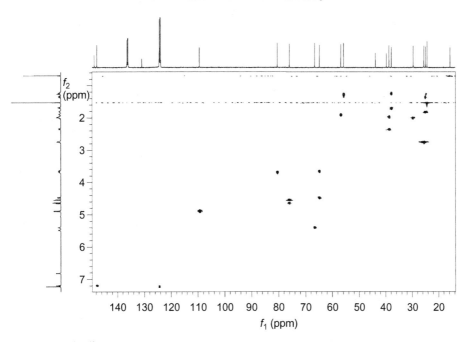

FIG. 2.6.5 The 2-D ^{1}H-^{13}C HSQC NMR spectrum of andrographolide in pyridine-d_5.

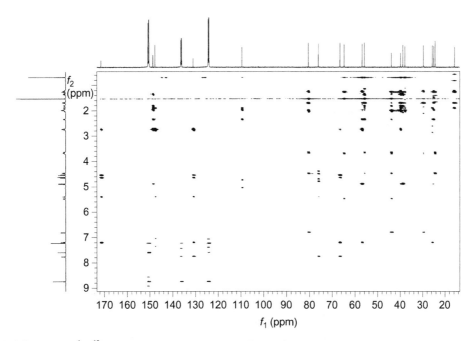

FIG. 2.6.6 The 2-D ^{1}H-^{13}C gHMBC NMR spectrum of andrographolide in pyridine-d_5.

TABLE 2.6.1 Predicted Multiplicities for the ^{1}H's of Andrographolide

Site in Molecule	Expected Multiplicity
1	$2 \times d^3$
2	$2 \times d^4$
3	d^2 (OH s)
5	d^2
6	$2 \times d^4$
7	$2 \times d^3$
9	t
11	$2 \times dt$
12	t
14	$2 \times d$
17	$2 \times d^2$
18	d^2 (OH s)
19	s
20	$2 \times d$ (OH s)
21	s

TABLE 2.6.2 Carbons of Andrographolide

Type of Carbon	Site Number
CH_3 (methyl)	19, 21
CH_2 (methylene)	1, 2, 6, 7, 11, 14, 17, 20
CH (methine)	3, 5, 9, 12, 18
C_{np} (nonprotonated)	4, 8, 10, 13, 15

Note that site 16 is an oxygen atom.

For the assignment of the ^{1}H and ^{13}C resonances to the hydrogen and carbon sites of andrographolide, we begin by identifying entry points. The lactone carbonyl ^{13}C signal (C15) is observed at 171.3 ppm—more than 20 ppm downfield from the next solute signal. The solvent, pyridine-d_5, generates three 1:1:1 triplet signals in our ^{13}C 1-D NMR spectrum; one of these triplets is the next closest ^{13}C signal to our carbonyl ^{13}C signal.

Searching for other, relatively easy assignments, we should look for other downfield resonances, which in this case will be the methine HSQC cross peak between the signals of H12 and C12 and the methylene cross peak between those of H14 and C14. The most downfield methine signal is found at $\delta_H = 7.20$ ppm and $\delta_C = 147.5$ ppm. Normally we can easily locate methylene signals on our HSQC spectrum because, for interesting molecules with their concomitant chiral centers, we find two ^{1}H signals correlating with one ^{13}C signal. But

TABLE 2.6.3 ^{1}H and ^{13}C NMR Signals of Andrographolide Listed by Group Type

^{1}H Signal (ppm)	^{13}C Signal (ppm)	Group Type
–	171.3	Nonprotonated
–	148.4	Nonprotonated
7.19	147.5	Methine
–	130.7	Nonprotonated
4.89 & 4.91	109.3	Methylene
3.69	80.3	Methine
4.53 & 4.63	75.9	Methylene
5.39	66.5	Methine
3.68 & 4.48	64.7	Methylene
1.89	56.8	Methine
1.27	55.8	Methine
–	43.8	Nonprotonated
–	39.7	Nonprotonated
1.96 & 2.35	38.7	Methylene
1.23 & 1.69	37.8	Methylene
1.99	29.5	Methylene
2.73	25.5	Methylene
1.36 & 1.81	24.9	Methylene
1.53	24.3	Methyl
0.69	15.7	Methyl

in some cases the ^{1}H signals from a methylene may be isochronous (have the same chemical shift), confounding our simple methods of locating two ^{1}H signals that correlate with one ^{13}C signal. While many HSQC spectra are collected in a DEPT-like phase-sensitive manner that allows one to distinguish between methine and methylene signals based on the sign of the cross peak, our black and white figures do not convey this information. We have other means of differentiating methine from methylene. Accounting for all of the ^{1}H signals in the ^{1}H 1-D spectrum requires we discover the signals from site 14, which is the downfield methylene signal at $\delta_H = 4.89$ & 4.91 ppm, $\delta_C = 109.3$ ppm. The assumption that the HSQC cross peaks near $\delta_H = 4.9$ and $\delta_C = 109$ ppm were the signals from two ^{1}H's bound to a ^{13}C was used to generate the relevant entries in Table 2.6.3, insofar as the row for the chemical shifts of 4.89 & 4.91, 109.3 ppm was labeled as a methylene group (it is remotely possible that this could be a methine plus a hydroxyl resonance—we can use the HSQC cross peak intensity as a proxy for cross peak volume if there is doubt. The methylene cross peak(s) should integrate to roughly twice the volume integral of the methine cross peaks). We should also note that the C12-C13 double bond is conjugated to the lactone carbonyl at site 15 and so we expect C12 to be downfield relative to benzene (it is, at 147.5 ppm, recall that benzene has a ^{13}C shift of 128.39 ppm).

Site 14 contains the only methylene group (carbon with two hydrogens) that generates signals in the downfield region, and so we write for site 14: δ_H = 4.89 & 4.91 ppm, and δ_C = 109.3 ppm.

In the 2-D ^{1}H-^{13}C HSQC NMR spectrum, we can identify as a group all of the methine and methylene signals arising from ^{1}H's on ^{13}C's directly bonded to oxygen: sites 3, 17, 18, and 20 comprise this group with a ^{1}H chemical shift range of 3.68–4.63 ppm and a ^{13}C chemical shift range of 64.7–80.3 ppm—we call this the midfield block of cross peaks of the HSQC spectrum. Recall that our standard for ^{1}H and ^{13}C signals alpha to the electronegative oxygen atom is for the ^{1}H chemical shift to be ~4 ppm and the ^{13}C shift to be ~60 ppm. Note that the midfield block includes the position of this important shift pair. It is important that we become adept at examining the overall appearance of the HSQC spectrum and recognizing, using our best fuzzy logic skills, general chemical shift trends manifested in how the various cross peaks are scattered across the HSQC spectrum. Moving from the midfield block of cross peaks along the pseudo-diagonal (a heteronuclear spectrum has no true diagonal), we next encounter two methine cross peaks, and farther along toward the origin (0,0) every signal stems from sites that are aliphatic. Please take a moment (unless you are cramming) to review the HSQC spectrum to observe the wonderful clarity afforded by the excellent chemical shift dispersion found in andrographolide—this molecule is a beautiful example of how easy a 20-carbon organic molecule can be to fully assign.

The methine resonance in our midrange block at δ_H = 5.39 ppm shares a cross peak in the 2-D ^{1}H-^{13}C gHMBC cross peak with the carbonyl carbon signal (171.3 ppm) at site 15, and so it is reasonable to attribute the generation of this cross peak to a $^3J_{HC}$ between H18 and C15, for the only other candidate is the H3 methine which is nine bonds distant. We therefore assign the H18 to 5.39 ppm and C18 to 66.5 ppm. With the 2-D ^{1}H-^{1}H COSY NMR spectrum, we can see a cross peak to a ^{1}H signal that does not reside on a carbon atom (at least not according to the HSQC spectrum) at 7.77 ppm. We assign this resonance to the site 18 hydroxyl ^{1}H resonance.

By process of elimination, we assign the other midrange methine signal at δ_H = 3.70 ppm and δ_C = 80.3 ppm to site 3. We can observe a COSY cross peak between the H3 signal at 3.70 ppm and the site 3 hydroxyl resonance at 6.81 ppm. Every time we assign a ^{1}H resonance from a hydrogen atom not on a carbon atom, we might, time permitting, consult the HSQC spectrum to verify that our ^{1}H signal shares no cross peak with a ^{13}C signal.

We now focus on the two methylenes generating HSQC cross peaks in the midrange chemical shift block. Again, the unique chemical shift of the carbonyl ^{13}C signal allows us to use the gHMBC spectrum to identify the H17 signals. We observe gHMBC cross peaks between the methylene ^{1}H signals at 4.53 & 4.63 ppm and the carbonyl ^{13}C signal (C15) at 171.3 ppm. The H20's are implausibly far from the carbonyl carbon (nine bonds) and so we assign, with a high degree of confidence, the site 17 shifts as follows: δ_H = 4.53 & 4.63 ppm and δ_C = 75.9 ppm. Recall that we use the row in Table 2.6.3 with the ^{1}H shifts of 4.53 & 4.63 ppm to find the corresponding ^{13}C shift for our methylene group.

Again, by process of elimination, we assign the remaining midfield methylene signals to site 20: δ_H = 3.68 & 4.48 ppm and δ_C = 64.7 ppm. In the COSY spectrum we observe a correlation between the H20 signals and a ^{1}H signal at 5.48 ppm. This signal arises from a ^{1}H not bound to carbon, i.e., the site 20 hydroxyl resonance.

Further use of the uniquely downfield carbonyl carbon signal allows us to assign the H11 signals, as we have already identified the signals from all more proximate protonated species. The ^{1}H signal at 2.73 ppm shares a gHMBC cross peak to the C15 carbonyl signal at 171.3 ppm, and so we can assign this signal as that of one of the two H11's. Lack of overlap at 2.73 ppm assures us that the cross peak cannot be caused by the signal of a ^{1}H other than one or both of the H11's, and so we can write for site 11 that $\delta_H = 2.73$ ppm and $\delta_C = 25.5$ ppm.

Sites 8 and 13 are the only two nonprotonated sites with sp^2-hybridized carbon atoms. If we examine the left edge of the gHMBC spectrum, we can readily identify the strong pair of cross peaks between the H17 signals and the C15 carbonyl signal. This same pair of ^{1}H signals (4.53 & 4.63 ppm) also shares cross peaks in the gHMBC spectrum with the nonprotonated C13 signal at 130.7 ppm (we verify in the HSQC spectrum that the ^{13}C signal at 130.7 ppm correlates with no ^{1}H signals).

We now use the H14 signal to identify the signals of the nonprotonated C8, the C7 methylene group, and the C9 methine. The H14 signals at 4.89 & 4.91 ppm share a gHMBC cross peak with a downfield nonprotonated carbon signal at 148.4 ppm which can only arise from site 8. The next nearest nonprotonated carbon is the sp^3-hybridized C10 which is four bonds distant, and C13, the only plausible alternative has already been assigned to the signal at 130.7 ppm.

The H11 signals (2.73 ppm) and the H14 signals (4.89 & 4.91 ppm) both share gHMBC cross peaks with a methine ^{13}C signal at 56.8 ppm. The methine signal cannot be that from site 5 (C5 is four and five bonds distant from the H11's and H14's, respectively), but instead must be from site 9 (C9 is two and three bonds distant from the H11's and H14's, respectively) and so we assign site 9 as follows: $\delta_H = 1.90$ ppm and $\delta_C = 56.8$ ppm.

The H14 signals at 4.89 & 4.91 ppm share strong gHMBC cross peak intensity with a methylene ^{13}C signal at 38.7 ppm. If the H14 signals correlate with only one methylene ^{13}C signal, we can reasonably expect that the correlation is with the signal of C7 and not with that of C6. We therefore assign the methylene group of site 7 as follows: $\delta_H = 1.97$ & 2.36 ppm and $\delta_C = 38.7$ ppm.

We can distinguish the signals of the two methyl groups at sites 19 and 21 by recognizing that the H9 signal should share a gHMBC cross peak with the C19 signal and the H20 signals should share gHMBC cross peaks with the C21 signal. We observe a cross peak between the H9 signal at 1.90 ppm and the methyl ^{13}C signal at 15.7 ppm, which we assign to site 19: $\delta_H = 0.69$ ppm and $\delta_C = 15.7$ ppm. We also observe gHMBC correlations between the H20 signals at 3.68 & 4.48 ppm and a methyl ^{13}C signal at 24.3 ppm, therefore allowing us to assign site 21 as follows: $\delta_H = 1.53$ ppm and $\delta_C = 24.3$ ppm.

We have two remaining nonprotonated ^{13}C signals at 39.7 and 43.8 ppm. These signals must arise from C4 and C10 because we have already identified the signals of the other three nonprotonated carbons: the carbonyl carbon (C15) and the two nonprotonated, sp^2-hybridized carbons (C8 and C13). The H19 signals at 0.69 ppm correlate with a nonprotonated ^{13}C signal at 39.7 ppm and so this signal is assigned to C10. Similarly, the methyl ^{1}H signals of site 21 at 1.53 ppm share a strong gHMBC correlation with the nonprotonated ^{13}C signal at 43.8 ppm, which we therefore assign to site 4.

The last remaining methine group at site 5 must have signal shifts as follows: $\delta_H = 1.27$ ppm and $\delta_C = 55.8$ ppm. Note that the H5 resonance at 1.27 ppm shares strong gHMBC cross peaks with both the signals of C4 (43.8 ppm) and C10 (39.7 ppm), thus confirming our assignment.

It now remains for us to assign the three remaining methylene groups. The H3 signal at 3.69 ppm shares COSY cross peaks with methylene ^{1}H signals at 1.99 ppm, allowing us to assign site 2 as follows: $\delta_H = 1.99$ ppm and $\delta_C = 29.5$ ppm. The H2 signals share COSY cross peaks with the methylene ^{1}H signals at 1.23 & 1.69 ppm, allowing us to assign site 1's signals as $\delta_H = 1.23$ & 1.69 ppm and $\delta_C = 37.8$ ppm.

By elimination, we can assign the site 6 methylene signals to $\delta_H = 1.36$ & 1.81 ppm and $\delta_C = 24.9$ ppm. We can confirm that our H6 assignment is correct by noting that the gHMBC spectrum contains a strong cross peak between the H6 signal at 1.81 ppm and the C8 signal at 148.4 ppm.

2.7 DEXAMETHASONE IN PYRIDINE-D_5

Dexamethasone has 22 carbons, five oxygens, four rings, and even a fluorine atom. The empirical formula of dexamethasone is $C_{22}H_{29}O_5F$. Two of the five oxygens are present as ketone carbonyls and one of these carbonyls is doubly conjugated, providing electron density to move the chemical shift of this carbonyl ^{13}C to a value lower than 200 ppm. Three hydroxyl groups account for the remaining oxygen atoms. The fluorine atom resides in an axial position at C9, at the junction of two of the rings. The structure of dexamethasone is shown in Fig. 2.7.1.

A small quantity of dexamethasone was dissolved in pyridine-d_5. This sample generated the spectra that appear in this section. The 1-D ^{1}H NMR spectrum of dexamethasone in pyridine-d_5 appears in Fig. 2.7.2. Fig. 2.7.3 contains the 1-D ^{13}C NMR spectrum of dexamethasone. The 2-D ^{1}H-^{1}H COSY NMR spectrum of dexamethasone is found in Fig. 2.7.4. Fig. 2.7.5 shows the 2-D ^{1}H-^{13}C HSQC NMR spectrum of dexamethasone. The 2-D ^{1}H-^{13}C gHMBC NMR spectrum of dexamethasone is shown in Fig. 2.7.6.

Examination of the structure of dexamethasone allows us to fill in an expected ^{1}H multiplicity table (Table 2.7.1) and a table of carbon types (Table 2.7.2). Examination of the 2-D ^{1}H-^{13}C HSQC NMR spectrum allows us to pair the signals of ^{1}H's with those of ^{13}C's one bond distant. Careful use of the HSQC and the integrals found in the 1-D ^{1}H NMR spectrum allow us to decide what signals are from methine groups versus from methylene groups and label

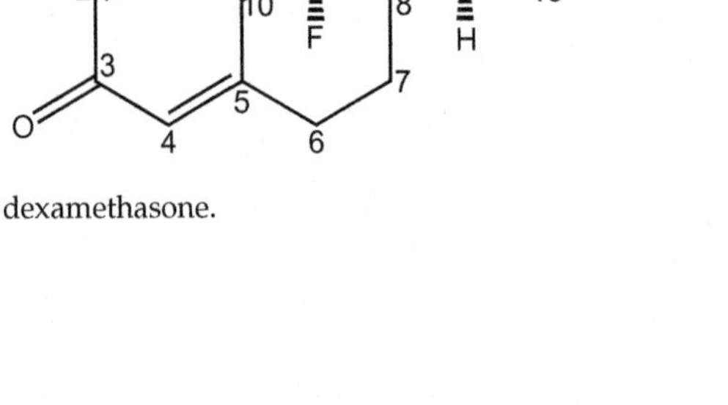

FIG. 2.7.1   The structure of dexamethasone.

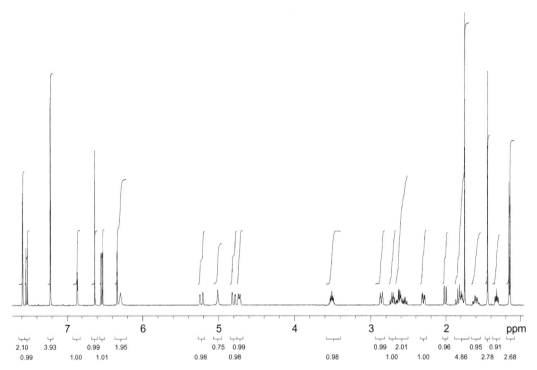

FIG. 2.7.2 The 1-D ^{1}H NMR spectrum of dexamethasone in pyridine-d_5.

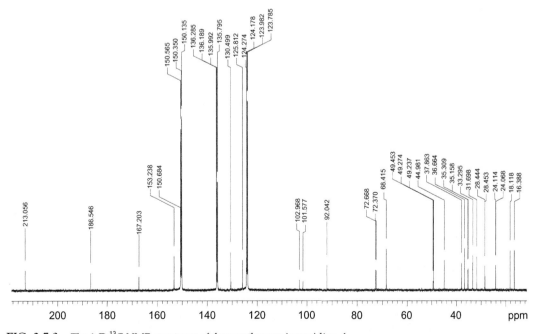

FIG. 2.7.3 The 1-D ^{13}C NMR spectrum of dexamethasone in pyridine-d_5.

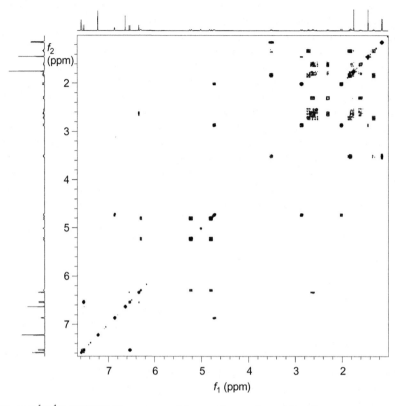

FIG. 2.7.4   The 2-D ^{1}H-^{1}H COSY NMR spectrum of dexamethasone in pyridine-d_5.

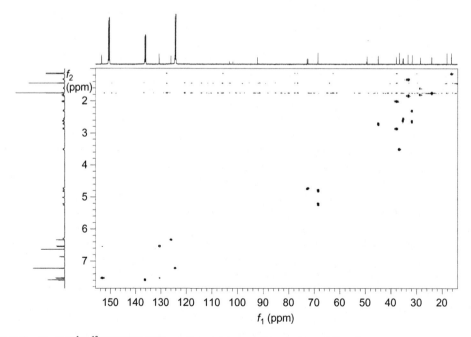

FIG. 2.7.5   The 2-D ^{1}H-^{13}C HSQC NMR spectrum of dexamethasone in pyridine-d_5.

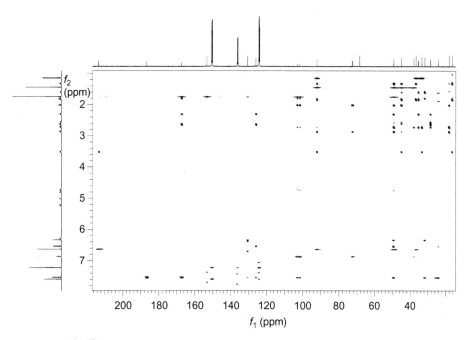

FIG. 2.7.6 The 2-D ^{1}H-^{13}C gHMBC NMR spectrum of dexamethasone in pyridine-d_5.

TABLE 2.7.1 Predicted Multiplicities for the ^{1}H's of Dexamethasone

Site in Molecule	Expected Multiplicity
1	d
2	d
4	s
6	td, dt
7	qd, dq
8	td^{2a}
11	d^{3a} (OH s)
12	t, d^2
14	td
15	td, dt
16	qt
17	(OH s)
18	s
19	s
20	d
22	2×d (OH s)

[a]The ^{19}F at site 9 adds an additional coupling.

2. CORTICOSTEROIDS

TABLE 2.7.2 Carbons of Dexamethasone

Type of Carbon	Site Number
CH_3 (methyl)	18, 19, 20
CH_2 (methylene)	6, 7, 12, 15, 22
CH (methine)	1, 2, 4, 8^a, 11^a, 14, 16
C_{np} (nonprotonated)	3, 5, 9^b, 10^a, 13, 17, 21

[a]*May be split by* [19]*F.*
[b]*Will be split by attached* [19]*F.*

them accordingly. This information appears in Table 2.7.3. Dexamethasone is a relatively rigid molecule. If we build the model, we will only find flexibility near C15-C17 and about the C17-C21 and C21-C22 single bonds. All six-membered rings are expected to be in the chair conformation.

Before we begin, it behooves us to go over the various annotations ([a], [b]) in the previous tables. The ^{19}F on C9 splits not just the C9 signal but also those of many other nearby spins as well. In these cases, the ^{13}C signal shifts are reported to the hundredth of a ppm to allow a better appreciation of how the magnitude of the $^nJ_{FC}$'s fall off as n increases (typically we only report ^{13}C chemical shifts to the nearest tenth of a ppm). Not only will some of our ^{1}H signal multiplets become more complicated, but also many of our ^{13}C signals will be split. We collected our 1-D ^{1}H and ^{13}C NMR spectra without ^{19}F decoupling so both J_{FC}'s and J_{FH}'s are expected to split the signals we observe. We must not lose our ability to account for each and every ^{13}C resonance. Care must be taken to ensure that a ^{19}F-split ^{13}C signal *will not* erroneously be interpreted as two distinct carbons (e.g., a split C9 signal could be mistaken for the signals of both C9 and C10). Examination of the gHMBC spectrum may help us locate multiple ^{13}C signals that share cross peaks with the same set of ^{1}H signals. By grouping these similar ^{13}C columns (f_1 is horizontal), we can avoid this error and preserve the integrity of our accounting methodology (pairing sites 1:1 to observed resonances).

We begin with the recognition that only one of the three methyl groups (site 20) is bound to a methine group (site 16), meaning that the H16 will split the H20 signal into a doublet. Site 20 is therefore assigned as $\delta_H = 1.15$ ppm and $\delta_C = 16.4$ ppm.

Inspection of the 1-D ^{13}C NMR spectrum allows us to assign site 21 as $\delta_C = 213.1$ ppm owing to its unique chemical environment as the only unconjugated ketone carbonyl. C3 is a doubly conjugated ketone and hence its signal is found at a chemical shift of less than 200 ppm: for site 3, $\delta_C = 186.5$ ppm.

The 2-D ^{1}H-^{1}H COSY contains cross peaks between the H20 signal at 1.15 ppm and a methine ^{1}H signal at 3.51 ppm, which must be that of H16. We assign site 16 as $\delta_H = 3.51$ ppm and $\delta_C = 36.7$ ppm.

In the COSY spectrum the H16 signal at 3.51 ppm shares a strong cross peak with a ^{1}H signal at 1.83 ppm which can only be the signal of one of the H15's because C17 on the other side of C16 is nonprotonated. This resonance overlaps to a small extent with another methylene ^{1}H signal at a slightly lower chemical shift. If we examine the HSQC spectrum, we can better appreciate this distinction and we can even obtain overlap free resonance locations (chemical shifts) for our ^{1}H dimension signals. We assign site 15 as $\delta_H = 1.33$ & 1.83 ppm and $\delta_C = 33.3$ ppm.

TABLE 2.7.3 ^{1}H and ^{13}C NMR Signals of Dexamethasone Listed by Group Type

^{1}H Signal (ppm)	^{13}C Signal (ppm)	Group Type
–	213.1	Nonprotonated
–	186.5	Nonprotonated
–	167.2	Nonprotonated
7.54	153.2	Methine
6.54	130.5	Methine
6.34	125.8	Methine
–	102.97/101.58[a]	Nonprotonated
–	92.0	Nonprotonated
–	72.67/72.37[a]	Methine
4.82 & 5.82	68.4	Methylene
–	49.45/49.27[a]	Nonprotonated
–	49.24	Nonprotonated
2.72	45.0	Methine
2.01 & 2.86	37.9	Methylene
3.51	36.7	Methine
2.54	35.31/35.16[a]	Methine
1.33 & 1.83	33.3	Methylene
2.31 & 2.64	31.7	Methylene
1.60 & 1.79	28.44/28.45[a]	Methylene
1.75	24.07/24.11[a]	Methyl
1.44	18.1	Methyl
1.15	16.4	Methyl

[a]Split by attached ^{19}F.

We can be confident that our H16 signal at 3.51 ppm couples with the ^{1}H shift at 1.83 ppm and not with the shift at 1.79 ppm if we examine the COSY spectrum. In the chemical shift range of 1.74–1.88 ppm our 1-D ^{1}H NMR spectrum contains five ^{1}H signals, three of which come from a methyl group that is a sharp singlet. If we begin by looking above the diagonal to the spot where δ_{f_1} = 3.51 ppm and δ_{f_2} =~ 1.8, the next cross peak we encounter as we move to the right (toward the diagonal) is clearly offset to the lower left relative to the other cross peak that is similarly disperse owing to coupling (proving it is not the nearby methyl ^{1}H signal which is a sharp singlet).

Using this visual comparison tool, we can save time and avoid having to measure the same shift multiple times. As software improves this methodology may appear quaint to future generations; however, it is important to guard against full digitization. Paper, especially the acid-free variety, still has a place in the world and may yet prove to be more enduring than any digital device. Besides, using pencil and paper (and maybe a ruler) to solve an NMR problem begets a certain understanding found satisfying by some.

The most downfield ^{13}C signal to which the H20 signals correlate in the 2-D ^{1}H-^{13}C gHMBC spectrum is the nonprotonated ^{13}C signal at 92.0 ppm. Of the three nonprotonated ^{13}C's within coupling range of the H20's, one is the already assigned carbonyl C21 and one is the aliphatic C13 whose signal shift is expected to be below 60 ppm. The 92.0 ppm shift is most consistent with the closest (3 bonds versus 4) nonprotonated ^{13}C. C17 has an alpha oxygen and an alpha ketone carbonyl, consistent with a signal shift over 60 ppm and so we write for site 17: $\delta_C = 92.0$ ppm.

We are able to assign the site 17 hydroxyl resonance using the gHMBC spectrum. The ^{1}H signal at 6.64 ppm shares gHMBC cross peaks with the C21 signal at 213.1 ppm, with the C17 signal at 92.0 ppm, with the C16 signal at 36.7 ppm, and with a nonprotonated ^{13}C signal at 49.24 ppm (f_1 resolution is coarse, but still is sufficient to show that the gHMBC cross peak is the more upfield the two ^{13}C resonances near 49 ppm). We assign the ^{13}C signal at 49.24 ppm as C13. For site 13, $\delta_C = 49.24$ ppm. The next nearest unassigned and nonprotonated ^{13}C to the hydroxyl group on C17 is C9, which is six bonds distant and expected to be split dramatically by its attached ^{19}F. You may have already identified C9 in Table 2.7.3 as the ^{13}C shift pair with the largest splitting, corresponding to the $^1J_{FC}$ (row 7 of Table 2.7.3) which is $\Delta\delta_C = 1.38$ ppm. Because 1 ppm = 125.6 Hz for ^{13}C on a 500 MHz instrument, the $^1J_{FC} = 175$ Hz (!)—we can write for site 9: $\delta_C = 102.3$ ppm (the average of 102.968 and 101.577, rounded to the nearest 10th).

Having already identified C21 as generating the most downfield ^{13}C signal at 213.1 ppm, we might attempt to identify the H22 methylene signals using the gHMBC spectrum, but these cross peaks are absent, likely because a plotting threshold was set too high when plotting the gHMBC spectrum. Because the H22's are bound to a carbon that is in turn bonded to oxygen, we can examine the midrange region of the HSQC spectrum and observe that only one methylene group populates the center of the HSQC spectrum near $\delta_H = 4$ ppm and $\delta_C = 60$ ppm. We write for site 22: $\delta_H = 4.82$ & 5.82 ppm and $\delta_C = 68.4$ ppm.

Moving from the upper right portion of the molecule to the lower left, we can exploit the combination of a strong *cis* $^3J_{HH}$ between H1 and H2 and the fact that both ^{1}H's are bound to sp^2-hybridized carbons. Because of conjugation, we know H1/C1 signal position will be downfield relative to the signal position of H2/C2. We observe a strong COSY cross peak pair between the ^{1}H signals at 6.54 and 7.54 ppm. We assign site 1 as $\delta_H = 7.54$ ppm and $\delta_C = 153.2$ ppm and for site 2, $\delta_H = 6.54$ ppm and $\delta_C = 130.5$ ppm.

By process of elimination we assign the last downfield ^{1}H signal at 6.34 ppm to H4. This is the last ^{1}H on an sp^2-hybridized carbon atom, and we observe that the H4 signal is indeed a singlet as predicted in Table 2.7.1. We can write for site 4: $\delta_H = 6.34$ ppm and $\delta_C = 125.8$ ppm.

C5 generates the last downfield ^{13}C signal now that we have assigned the signals of C21, C3, and C1. The nonprotonated ^{13}C resonance at 167.2 ppm shares a strong gHMBC cross peak with the H1 resonance, thus confirming that our assignment at site 5 can be written as $\delta_C = 167.2$ ppm. Also note that H1 has the same bond geometry (*trans*-$^3J_{HC}$) to the C3 carbonyl ^{13}C, and making the cross peak between the signals of H1 and C3 similarly intense in the gHMBC spectrum.

One of the two methyl ^{1}H singlets (at 1.75 ppm) that we have yet to assign correlates strongly to the C5 signal (167.2 ppm) in the gHMBC spectrum, and so the ^{1}H signal at 1.75 ppm must be that of H19, not that of H18, for the H19's are 3 bonds from C5 while the H18's are seven bonds distant. We assign site 19 signals as $\delta_H = 1.75$ ppm and $\delta_C = 24.1$ ppm.

By elimination we can assign site 18 as $\delta_H = 1.44$ ppm, and $\delta_C = 18.1$ ppm. The H18 signals share two confirming gHMBC cross peaks with the C13 signal at 49.24 ppm and with the C17 signal at 92.0 ppm. The H18 signals also share gHMBC cross peaks with two unassigned ^{13}C signals: the first is a methylene ^{13}C signal at 37.9 ppm, and the second is a methine ^{13}C signal at 45.0 ppm. We assign the methylene signal near site 18 to site 12: $\delta_H = 2.01$ & 2.86 ppm and $\delta_C = 37.9$ ppm. We assign the methine signal near site 18 to site 14: $\delta_H = 2.72$ ppm and $\delta_C = 45.0$ ppm.

The H19 signal at 1.75 ppm shares a gHMBC cross peak with one or both of the ^{13}C signals near 49 ppm—probably the more downfield of the two. Because we have already assigned the upfield resonance to C13 and because the downfield resonance is split into two peaks by the ^{19}F attached to C9, we assign the downfield ^{13}C signal to C10, writing for site 10: $\delta_C = 49.36$ ppm or 49.4 ppm (we average the two C10 signals).

At this point we discuss how ^{19}F is splitting our ^{13}C signals. We expect the C9 signal to be split, but also probably those of C8, C10, and C11. The nonprotonated ^{13}C resonance most split by ^{19}F is observed at 102.97 and 101.58 ppm, a difference of 175 Hz which can only arise from the $^1J_{FC}$. Additionally, we see that the C19 signal is split (at 24.1 ppm), and even one methylene ^{13}C signal features a small splitting, which presumably is the signal from site 7 (28.4 ppm). We can therefore write for site 7: $\delta_H = 1.60$ & 1.79 ppm and $\delta_C = 28.4$ ppm. Examination of the ^{1}H resonance at 1.60 ppm shows that the multiplet is a quartet of doublets, consistent with our expectation that the axial ^{1}H on C7 will couple strongly via *trans* $^3J_{HH}$'s with the equatorial H7 (H7$_{eq}$) as well as with the neighboring axial ^{1}H's on C6 and C8 (H6$_{ax}$ and H8$_{ax}$). These three large couplings impart the apparent quartet pattern to the axial H7 signal, and the small gauche coupling that results from the H7$_{ax}$ to H6$_{eq}$ contributes the further doubling of each of the quartet's lines.

By elimination we can assign the last methylene group to site 6: $\delta_H = 2.31$ & 2.64 ppm and $\delta_C = 31.7$ ppm.

We now must differentiate between our last two methines at sites 8 and 11. Because C11 has an attached oxygen atom, we can easily determine that the H11/C11 signal cross peak will appear at a chemical shift position that is more downfield in the HSQC spectrum. We assign site 11 as follows: $\delta_H = 4.73$ ppm and $\delta_C = 72.5$ ppm. We assign site 8 as follows: $\delta_H = 2.54$ ppm and $\delta_C = 35.2$ ppm.

2.8 (−)-NORGESTREL IN PYRIDINE-D₅

(−)-Norgestrel has 21 carbons, two oxygens, four rings, a pendant ethyl group, and a triply bonded terminal alkyne group. The empirical formula of norgestrel is $C_{21}H_{26}O_2$. With nine methylene groups, we anticipate that our assignment of the resonances to the hydrogen and carbon sites of norgestrel will dwell upon these groups more than any others in the molecule. One of the two oxygens is found in a conjugated ketone carbonyl and the second in a hydroxyl group. The structure of (−)-norgestrel is shown in Fig. 2.8.1.

A small amount of (−)-norgestrel was dissolved in pyridine-d_5. This sample was used to obtain the spectra appearing in this section. The 1-D ^{1}H NMR spectrum of

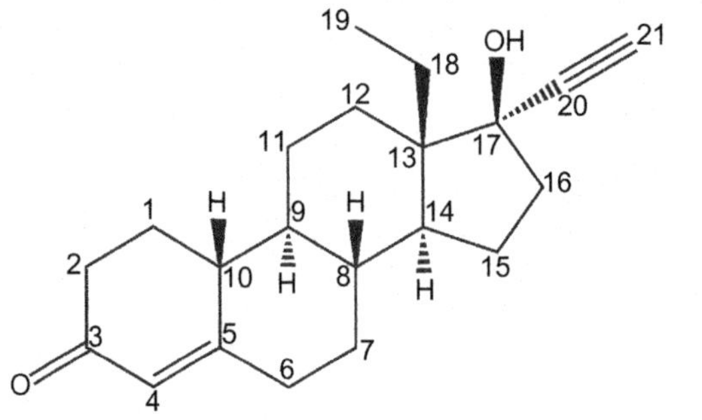

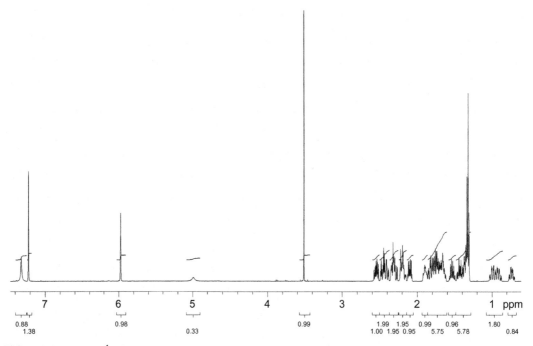

FIG. 2.8.1 The structure of (−)-norgestrel.

FIG. 2.8.2 The 1-D ^{1}H NMR spectrum of (−)-norgestrel in pyridine-d_5.

(−)-norgestrel in pyridine-d_5 is found in Fig. 2.8.2. An expanded portion of the 1-D ^{1}H NMR spectrum of (−)-norgestrel appears in Fig. 2.8.3. Fig. 2.8.4 contains the 1-D ^{13}C NMR spectrum of (−)-norgestrel. The 2-D ^{1}H-^{1}H COSY NMR spectrum of (−)-norgestrel is shown in Fig. 2.8.5, while an expanded portion of the COSY spectrum appears in Fig. 2.8.6. Fig. 2.8.7 shows the 2-D ^{1}H-^{13}C HSQC NMR spectrum of (−)-norgestrel, and an expanded portion of the HSQC spectrum is shown in Fig. 2.8.8. The 2-D ^{1}H-^{13}C gHMBC NMR spectrum of (−)-norgestrel is shown in Fig. 2.8.9. An expanded portion of the gHMBC spectrum of (−)-norgestrel appears in Fig. 2.8.10.

Examination of the structure of (−)-norgestrel allows us to fill in an expected ^{1}H multiplicity table (Table 2.8.1) and a table of carbon types (Table 2.8.2). Information found in the

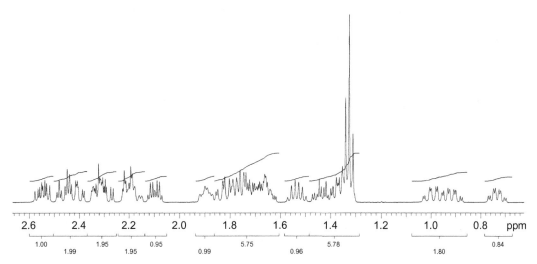

FIG. 2.8.3 An expanded portion of the 1-D ^{1}H NMR spectrum of (−)-norgestrel in pyridine-d_5.

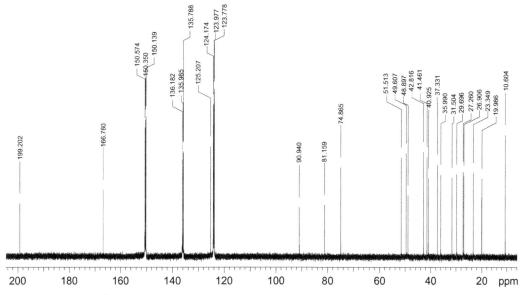

FIG. 2.8.4 The 1-D ^{13}C NMR spectrum of (−)-norgestrel in pyridine-d_5.

2-D ^{1}H-^{13}C HSQC NMR spectrum allows grouping of shifts of ^{1}H's bound to a common ^{13}C. A combination of 1-D ^{1}H integrals and cross peaks in the 2-D ^{1}H-^{13}C HSQC NMR spectrum allow the determination of which signals stem from methine, methylene, or methyl groups. We label them accordingly. We expect that the terminal alkyne ^{1}H and ^{13}C signals may not generate a satisfactory cross peak in the HSQC spectrum because the $^1J_{CH}$ between H21 and C21 is expected to be around 250 Hz, far greater than the expectation value of 140 Hz used to calculate a delay used in the HSQC pulse sequence. That is, the terminal alkyne ^{1}H-^{13}C

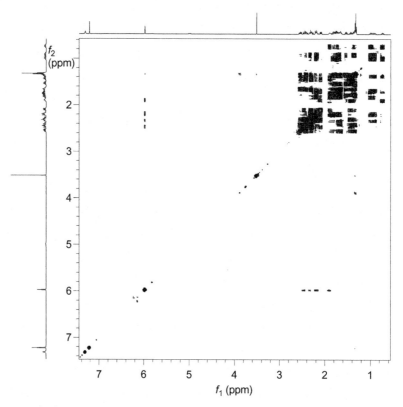

FIG. 2.8.5   The 2-D ^{1}H-^{1}H COSY NMR spectrum of (−)-norgestrel in pyridine-d_5.

signal cross peak may be absent from the HSQC spectrum. However, we should be able to use information found in the 2-D ^{1}H-^{13}C gHMBC spectrum to identify the ^{1}H and ^{13}C shifts for this site. Information relating ^{1}H signals to those of their attached ^{13}C's appears in Table 2.8.3. (−)-Norgestrel has very little flexibility, with only limited conformational variability available at the C13-C18 bond (the ethyl group can spin around) and near C15-C17. All six-membered rings are expected to be in the chair conformation whenever possible, but sp^2-hybridization of some of the carbons in the lower left two rings will flatten these two rings.

Our entry into the assignment of (−)-norgestrel begins with the sole methyl group at site 19. From the 1-D ^{1}H NMR spectrum we observe the H19 signal is partitioned into the predicted triplet multiplet pattern with a chemical shift of 1.33 ppm, allowing us to write for site 19: $\delta_H = 1.33$ ppm and $\delta_C = 10.6$ ppm.

A second assignment we can make with a high degree of confidence is to assign the carbonyl carbon C3 to the signal that is the most downfield ^{13}C chemical shift observed, allowing us to write for site 3: $\delta_C = 199.2$ ppm.

Using our understanding of how a resonance structure can be written to place a negative formal charge on the carbonyl oxygen atom and a positive formal charge on the C5 carbon, we predict that the C5 signal will be the next most downfield ^{13}C resonance after that of the carbonyl ^{13}C. Accordingly, we write for site 5: $\delta_C = 166.8$ ppm.

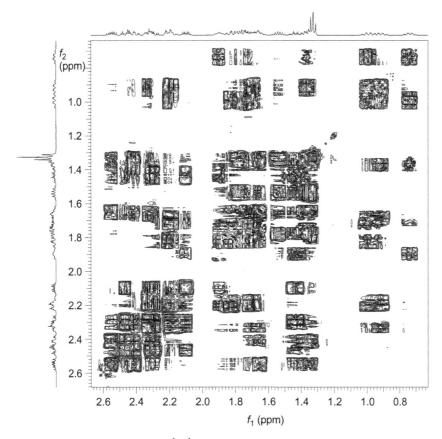

FIG. 2.8.6 An expanded portion of the 2-D ^{1}H-^{1}H COSY NMR spectrum of (−)-norgestrel in pyridine-d_5.

After marking the solvent resonances in the 2-D ^{1}H-^{13}C HSQC NMR spectrum, we are able to identify the sole remaining downfield HSQC cross peak corresponding to site 4: $\delta_H = 5.98$ ppm and $\delta_C = 125.2$ ppm.

We have many (eight) methylene groups to assign. We begin this task by using the proximity of sites 1 and 2 to the carbonyl ^{13}C at site 3 to locate the H1 and H2 signals. Because site 2 is alpha to the carbonyl group at site 3, we expect the H2/C2 chemical shifts to be further downfield than the H1/C1 shifts. Examination of the left edge (f_1 is horizontal) of the 2-D ^{1}H-^{13}C gHMBC NMR spectrum reveals four cross peaks shared by the C3 signal and the signals from the two pairs of H1's and H2's. The downfield signal pair is from the H2's at 2.31 & 2.47 ppm and the upfield signal pair is from the H1's at 1.45 & 2.11 ppm. We write for site 2: $\delta_H = 2.31$ & 2.47 ppm and $\delta_C = 37.3$ ppm and for site 1: $\delta_H = 1.45$ & 2.11 ppm and $\delta_C = 27.3$ ppm. We can confirm the rectitude of our assignment of the site 1 ^{1}H signals by noting that the doublet of quartets at 2.11 ppm (the H1$_{eq}$ resonance) correlates strongly with the C5 ^{13}C resonance at 166.8 ppm in the gHMBC spectrum.

Other gHMBC cross peaks correlating with the C5 signal allow us to identify ^{1}H signals from the methine group at site 10 and the methylene group at site 6. Both the methines H9 and

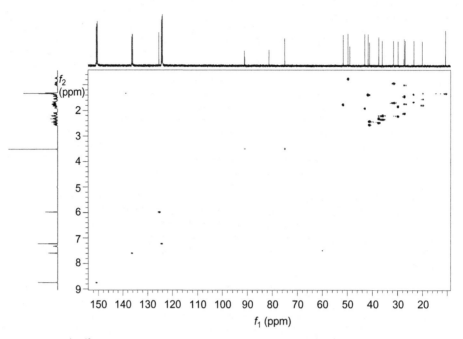

FIG. 2.8.7   The 2-D ^{1}H-^{13}C HSQC NMR spectrum of (−)-norgestrel in pyridine-d_5.

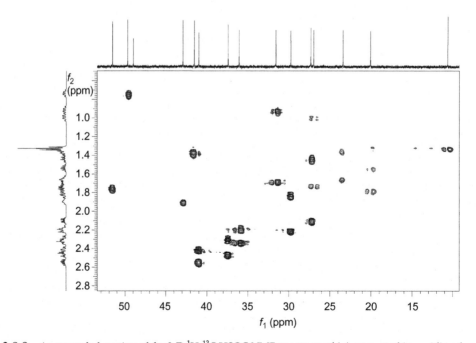

FIG. 2.8.8   An expanded portion of the 2-D ^{1}H-^{13}C HSQC NMR spectrum of (−)-norgestrel in pyridine-d_5.

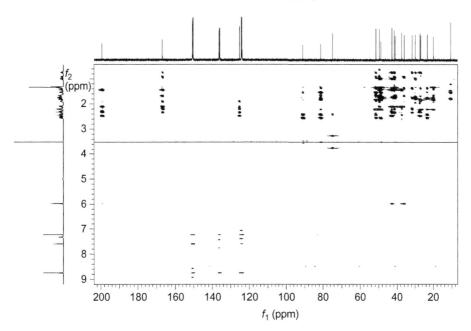

FIG. 2.8.9 The 2-D ^{1}H-^{13}C gHMBC NMR spectrum of (−)-norgestrel in pyridine-d_5.

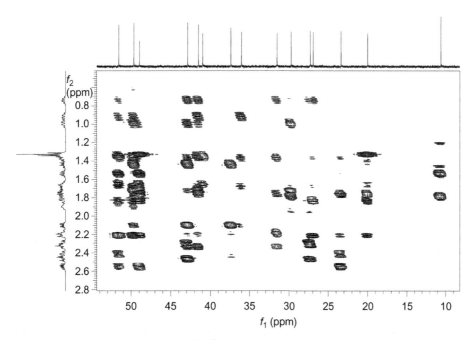

FIG. 2.8.10 An expanded portion of the 2-D ^{1}H-^{13}C gHMBC NMR spectrum of (−)-norgestrel in pyridine-d_5.

TABLE 2.8.1   Predicted Multiplicities for the ^{1}H's of (−)-Norgestrel

Site in Molecule	Expected Multiplicity
1	td, dt
2	qd, dq
4	s
6	td, dt
7	qd, dq
8	qd
9	qd
10	td
11	qd, dq
12	td, dt
14	td
15	qd, dq
16	td, dt
17	(OH s)
18	2×qd
19	t
21	s

TABLE 2.8.2   Carbons of (−)-Norgestrel

Type of Carbon	Site Number
CH_3 (methyl)	19
CH_2 (methylene)	1, 2, 6, 7, 11, 12, 15, 16, 18
CH (methine)	4, 8, 9, 10, 14, 21
C_{np} (nonprotonated)	3, 5, 13, 17, 20

H10 signals share cross peaks with the C5 signal in the gHMBC spectrum—we observe the methine ^{1}H signal at 0.75 ppm correlates with the C5 signal at 166.8 ppm as does the methine ^{1}H signal at 1.91 ppm. We differentiate between the H9 signal and H10 signal by noting that in the 1-D ^{1}H NMR spectrum, the resonance at 0.75 ppm is clearly resolved and is a beautiful example of a quartet of doublets which is only consistent with our prediction in Table 2.8.1 for the splitting pattern of the signal of H9 and not H10. Therefore we can write for site 9: $\delta_H = 0.75$ ppm and $\delta_C = 49.6$ ppm and for site 10: $\delta_H = 1.91$ ppm and $\delta_C = 42.8$ ppm.

The H4 signal shares its two strongest gHMBC cross peaks with the *trans*-$^3J_{CH}$-coupled C10 signal at 42.8 ppm and with the *trans*-$^3J_{CH}$-coupled C2 signal at 37.3 ppm. The H4 signal

TABLE 2.8.3 ^{1}H and ^{13}C NMR Signals of (−)-Norgestrel Listed by Group Type

^{1}H signal (ppm)	^{13}C Signal (ppm)	Group Type
–	199.2	Nonprotonated
–	166.8	Nonprotonated
5.98	125.2	Methine
–	90.9	Nonprotonated
–	81.2	Nonprotonated
3.51	74.9	Methine
1.76	51.5	Methine
0.75	49.6	Methine
–	48.9	Nonprotonated
1.91	42.8	Methine
1.38	41.5	Methine
2.42 & 2.55	40.9	Methylene
2.31 & 2.47	37.3	Methylene
2.20 & 2.34	36.0	Methylene
0.92 & 1.69	31.5	Methylene
1.84 & 2.22	29.7	Methylene
1.45 & 2.11	27.3	Methylene
1.00 & 1.73	26.9	Methylene
1.36 & 1.57	23.3	Methylene
1.55 & 1.79	20.0	Methylene
1.33	10.6	Methyl

shares its next most intense gHMBC cross peak with a methylene ^{13}C signal at 36.0 ppm which we take as the signal of the cis-$^3J_{CH}$-coupled C6 (we expect C6 to couple to H4 slightly less strongly than C2 and C10 according to the ^3J Karplus relationship). The methylene signals from the ^{1}H's on the ^{13}C whose shift is 36.0 ppm (at 2.20 & 2.32 ppm) share gHMBC cross peaks with the C5 signal at 166.8 ppm. These two pieces of evidence allow us to write for site 6: $\delta_H = 2.20$ & 2.32 ppm and $\delta_C = 36.0$ ppm.

The well-resolved ^{1}H resonance at 0.75 ppm from the methine H9 shares a strong COSY cross peak with a quartet of doublets at 1.00 ppm, which has been identified as a methylene ^{1}H resonance in row 18 (4th from the bottom) of Table 2.8.3. This quartet of doublets must be the signal of the axial H11 because the H8 signal, although it has the correct multiplicity, is from a methine and not a methylene, and also because H12$_{ax}$ is (1) too far away to couple so strongly to an axial ^{1}H and (2) is predicted to be a triplet of doublets, not a quartet of doublets. We assign site 11 as follows: $\delta_H = 1.00$ & 1.73 ppm and $\delta_C = 26.9$ ppm.

We can obtain another assignment from the unimpinged resolving of the H9 methine signal at 0.75 ppm. From our knowledge of the $^3J_{HH}$ Karplus relationship we expect that H9, which is axial, will couple strongly to H8, H10, and H11$_{ax}$ (the axial H11). We have already identified the signals of H10 and H11$_{ax}$, so the examination of the COSY spectrum allows us to discover that the H9 signal at 0.75 ppm also correlates strongly with a methine signal at 1.38 ppm. This must be the H8 signal and so we have found the signal of our third 1,2-diaxial ^3J coupling partner and we write for site 8: $\delta_H = 1.38$ ppm and $\delta_C = 41.5$ ppm.

Having identified the signals of both the H6's and H8, we focus on site 7. Because the 2-D ^{1}H-^{1}H COSY spectrum is nightmarishly crowded near 1.4, 1.7, and 2.4 ppm, we will use the gHMBC spectrum to use known ^{1}H shifts to identify unknown ^{13}C signals, and then look for agreement from the ^{1}H shifts of the newly identified ^{13}C (afforded via HSQC) in the 1-D ^{1}H spectrum, as well as in the COSY and the gHMBC spectra. If we have found the ^{13}C signal of the right site, then the multiplicities of the new ^{1}H signals will match—we may have to examine the COSY and/or the gHMBC spectra to assess multiplet structure if the 1-D ^{1}H spectrum is overlapped near the shift of interest. The ^{1}H signals of the new site should also share COSY cross peaks with the signals of nearby ^{1}H's that are already assigned. The ^{1}H shifts of the new site should additionally correlate in the gHMBC spectrum with known ^{13}C signals that are close (in terms of number of bonds distant) and in a favorable geometry to promote effective coupling through the overlap of the molecular orbitals.

The H6 signals at 2.20 & 2.34 ppm and the H8 signal at 1.38 ppm share gHMBC cross peaks with a methylene ^{13}C signal at 31.5 ppm. The ^{1}H signals of the methylene with a ^{13}C shift of 31.5 ppm are 0.92 & 1.69 ppm. The upfield multiplet is resolved well enough for us to discern that it is a quartet of doublets, consistent with the predicted multiplicity of the H7$_{ax}$ signal (the axial ^{1}H on site 7). In the COSY spectrum, the ^{1}H signals at 0.93 & 1.69 ppm both share significant cross peak intensity with the H8 signal at 1.38 ppm and with the H6 signals at 2.20 & 2.34 ppm. In the gHMBC spectrum, the 0.93 & 1.69 ^{1}H signals correlate with both the signals of C6 at 36.0 ppm and C8 at 41.5 ppm. Thus we are now comfortable writing for site 7: $\delta_H = 0.93$ & 1.69 ppm and $\delta_C = 31.5$ ppm.

Rather than use the HSQC spectrum to find the cross peak from the large $^1J_{CH}$ we expect for the alkynic methine group at site 21, it is easier to examine the gHMBC spectrum to find the coupled pair of HMQC cross peaks for the signals of the spins of site 21 in the midfield chemical shift range. A $^1J_{CH}$ of 250 Hz generates a half ppm splitting along the ^{1}H (f_2) axis of the gHMBC spectrum. The ^{1}H signal at 3.51 ppm shows this pronounced coupling to the ^{13}C signal at 74.9 ppm. Compare the clarity afforded using this approach to the ambiguity of the HSQC spectrum, which contains a weak cross peak between $\delta_H = 3.51$ ppm and $\delta_C = 74.9$ ppm, but also a second, misleading cross peak correlating with the ^{13}C signal at 90.9 ppm. We write for site 21: $\delta_H = 3.51$ ppm and $\delta_C = 74.9$ ppm.

There is now but one methine group whose signals are left to be identified. The site 14 ^{1}H/^{13}C pair is assigned as follows: $\delta_H = 1.76$ ppm and $\delta_C = 51.5$ ppm. The ^{1}H signal of H14 is not especially useful in confirming this assignment, as it falls into a heavily overlapped chemical shift region of the ^{1}H NMR spectrum. The ^{13}C signal of C14 does, however, share a gHMBC cross peak with the H8 signal at 1.38 ppm.

C17 and C20 are both nonprotonated carbons located at the top right of the molecule as it is pictured. The two ^{13}C signals at 90.9 and 81.2 ppm can be attributed to these two sites, but differentiating them is not easy because a profound t_1 ridge in the gHMBC spectrum prevents us from observing the cross peak between the signals of H21 and C20. We have two other clues as to the identity of these two ^{13}C resonances. One, we expect that C17 will couple to more ^{1}H's than will C20, because C20 is more distant from any nearby ^{1}H's, and two, we might expect that the $^2J_{CH}$ of perhaps 50 Hz that couples H21 and C20 may generate a weak cross peak in the HSQC spectrum between the H21 and C20 signals. In the gHMBC spectrum, the ^{13}C signal at 81.2 ppm shares seven discernible and well-resolved cross peaks, while the ^{13}C signal at 90.9 ppm participates in but three—this suggests that the more social C17 generates the signal we observe at 81.2 ppm and the more isolated C20 generates the signal at 90.9 ppm. In the HSQC spectrum, recall that we observed a misleading cross peak when we were assigning site 21. This cross peak correlated the H21 signal at 3.51 ppm with the ^{13}C signal at 90.9 ppm and is attributable to a very large $^2J_{CH}$ afforded by the uniquely linear nature of the sp-hybridized carbon atom at site 21. This weak cross peak in the HSQC spectrum suggests that the ^{13}C signal at 90.9 ppm is that of C20 and the ^{13}C signal at 81.2 ppm is that of C17. Chemical shift prediction software also assists in resolving this type of question, as long as ^{13}C chemical shift differences are greater than the uncertainty associated with the prediction. We write for site 20: $\delta_C = 90.9$ ppm and for site 17: $\delta_C = 81.2$ ppm.

The methylene group with ^{1}H shifts of 2.42 & 2.55 ppm share gHMBC cross peaks with the C20 signal at 90.9 ppm and with the C17 signal at 81.2 ppm. It is reasonable that these correlations are attributable to the signals of the H16's and not the H15's. The axial H15 is not expected to significantly couple to the C20 four bonds away, especially because the C20-C21 alkyne group occupies a pseudo-axial position and not the pseudo-equatorial position (we cannot speak of true axial and equatorial because the top-right ring has only five members). We tentatively write for site 16: $\delta_H = 2.42$ & 2.55 ppm and $\delta_C = 40.9$ ppm.

A set of methylene ^{1}H signals at 1.36 & 1.67 ppm correlate with the putative C signal at 40.9 ppm in the gHMBC spectrum. These signals can be ascribed to the site 15 ^{1}H's. The tentatively assigned 2.42 & 2.55 ppm ^{1}H signals of the H16's correlate strongly with the C15 signal at 23.3 ppm in the gHMBC spectrum, confirming the consistency of the site 15 and 16 methylene ^{1}H and ^{13}C signal assignments. Of limited utility but still somewhat of a confirmation, we observe a gHMBC cross peak shared between the H14 signal at 1.76 ppm and the C15 ^{13}C signal at 23.3 ppm. We assign site 15 as follows: $\delta_H = 1.36$ & 1.67 ppm and $\delta_C = 23.3$ ppm and accept the validity of the site 16 assignment in the previous paragraph.

The intense methyl ^{1}H triplet signal from site 19 at 1.33 ppm participates in two prominent cross peaks in the gHMBC spectrum with ^{13}C signals at 48.9 and 20.0 ppm. The 48.9 ppm ^{13}C signal is from a nonprotonated carbon site and the 20.0 ppm ^{13}C signal from a methylene group. Site 13 is the last of the nonprotonated sites, and so the ^{13}C signal at 48.9 ppm must be that of C13. We write for site 13: $\delta_C = 48.9$ ppm.

We have two remaining unassigned methylene groups at sites 12 and 18. We assign the methylene group whose ^{13}C resonance is found at 20.0 ppm to site 18 because this ^{13}C signal correlates strongly with the signals of the H19's (the COSY spectrum was not especially useful in finding the H18 signal's chemical shift using the H19 signals as the starting point owing to extensive overlap). We write for site 18: $\delta_H = 1.55$ & 1.79 ppm and $\delta_C = 20.0$ ppm.

By process of elimination we write for site 12: $\delta_H = 1.84$ & 2.22 ppm and $\delta_C = 29.7$ ppm. In the gHMBC spectrum we observe confirming correlations between the signals of the H12's at 1.84 & 2.22 ppm and ^{13}C signals at 81.2 (C17), 51.5 (C14), 49.6 (C9), 48.9 (C13), 26.9 (C11), and 20.0 ppm (C18).

2.9 DIGOXIGENIN IN ACETONE-D_6

Digoxigenin has 23 carbons, five rings, five oxygens, two of which are present in a conjugated lactone and three as hydroxyls, an IHD of seven, and two *cis* ring-junctions. Flexibility is limited to ring V rotation about the C17-C20 single bond and the fluxional nature of the two five-membered rings (rings IV and V). The empirical formula of digoxigenin is $C_{23}H_{34}O_5$. Just like (-)-norgestrel, digoxigenin has nine methylene groups that will likely dominate the demands placed on us as we assign the NMR signals of digoxigenin to their corresponding molecular sites. The structure of digoxigenin is shown in Fig. 2.9.1.

A small amount of digoxigenin was dissolved in acetone-d_6. This sample generated the spectra found in this section. The 1-D ^{1}H NMR spectrum of digoxigenin in acetone-d_6 is found in Fig. 2.9.2. Fig. 2.9.3 contains the 1-D ^{13}C NMR spectrum of digoxigenin. The upfield portion of the 1-D ^{13}C NMR spectrum of digoxigenin is found in Fig. 2.9.4 and the downfield portion of the 1-D ^{13}C NMR spectrum is found in Fig. 2.9.5. The 2-D ^{1}H-^{1}H COSY NMR spectrum of digoxigenin appears in Fig. 2.9.6. Fig. 2.9.7 shows the 2-D ^{1}H-^{13}C HSQC NMR spectrum of digoxigenin, and an expanded portion of the HSQC spectrum is shown in Fig. 2.9.8. The 2-D ^{1}H-^{13}C gHMBC NMR spectrum of digoxigenin appears in Fig. 2.9.9. An expanded portion of the gHMBC spectrum of digoxigenin is shown in Fig. 2.9.10.

We predict the expected multiplicities of the ^{1}H signals of digoxigenin by examining its structure and put our results into Table 2.9.1. We take the time to do this so that we will be in a position to quickly and easily refer to this table for multiplet confirmations whenever possible. Axial ^{1}H's on six-membered rings where there exists two or three strong couplings ($^2J_{HH}$ plus one or two *trans*-$^3J_{HH}$'s) will generate a signal that appears as a triplet or a quartet that may prove diagnostic. Based on examination of the structure of digoxigenin we also, for the purpose of accounting for all the NMR signals, fill in a table of carbon types (Table 2.9.2). Cross peak information found in the 2-D ^{1}H-^{13}C HSQC NMR spectrum allows grouping of signals of ^{1}H's bound to a common ^{13}C. By accounting for the stoichiometric representation of 1-D ^{1}H signal intensities (integrals) and combining this abundance information with information

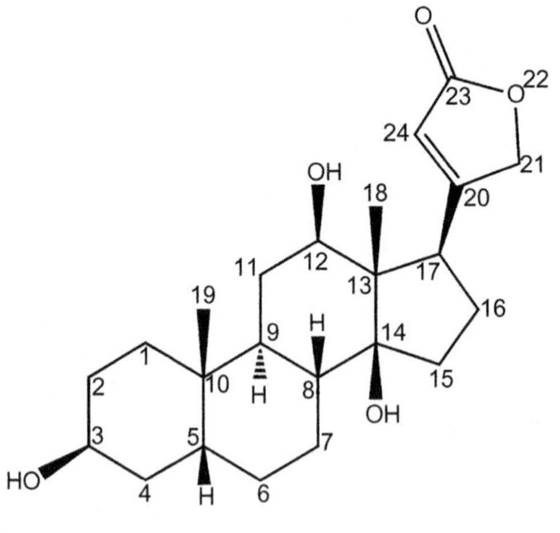

FIG. 2.9.1 The structure of digoxigenin.

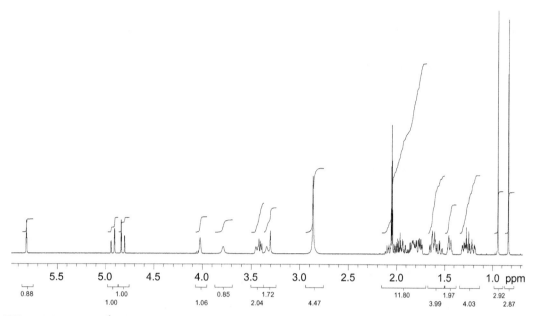

FIG. 2.9.2 The 1-D ^{1}H NMR spectrum of digoxigenin in acetone-d_6.

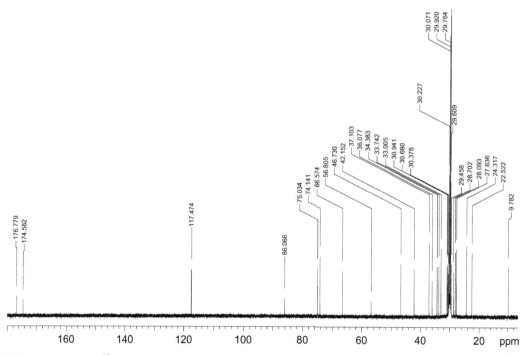

FIG. 2.9.3 The 1-D ^{13}C NMR spectrum of digoxigenin in acetone-d_6.

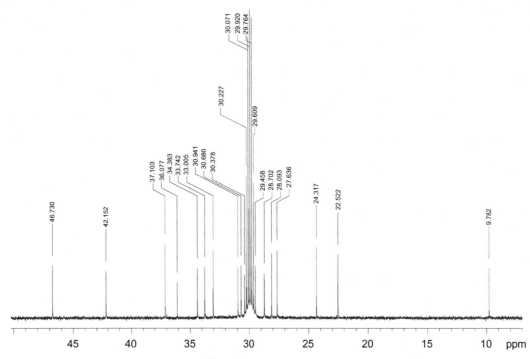

FIG. 2.9.4   The upfield portion of the 1-D ^{13}C NMR spectrum of digoxigenin in acetone-d_6.

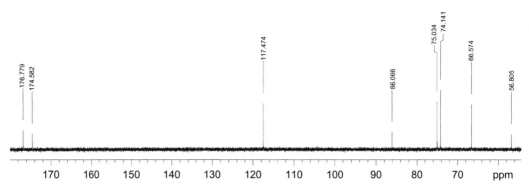

FIG. 2.9.5   The downfield portion of the 1-D ^{13}C NMR spectrum of digoxigenin in acetone-d_6.

contained in the cross peaks of the 2-D ^{1}H-^{13}C HSQC NMR spectrum, we are able to determine which signals arise from methine, methylene, and methyl groups. Nonprotonated ^{13}C signals are identified using the 1-D ^{13}C spectrum. We label each row in Table 2.9.3 accordingly. Not all of the six-membered rings will exist in the chair conformation because the fusion between rings I & II and also between III & IV (ring I is lower left) are *cis*.

Digoxigenin is a very challenging molecule to assign. The myriad methylene groups generate signals which densely populate several upfield regions of the ^{1}H chemical shift axis. There are also seven ^{13}C signals observed within four ppm of the solvent signal at 29.92 ppm.

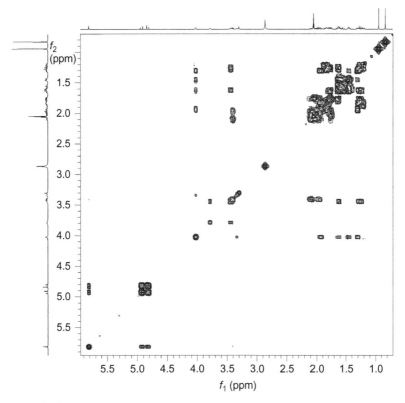

FIG. 2.9.6 The 2-D ^{1}H-^{1}H COSY NMR spectrum of digoxigenin in acetone-d_6.

Recall that in the ^{13}C spectrum the dominant solvent species we observe will be ^{13}C with three ^{2}H's attached and so we expect seven lines (each additional ^{2}H adds two lines) and in the ^{1}H spectrum the dominant ^{1}H-containing solvent species will be acetone-d_5,h_1 so the ^{1}H will couple to two ^{2}H's (on the same methyl group) and generate a signal with five lines.

Digoxigenin has only one ^{1}H bound to an sp^2-hybridized carbon atom. H24 can be unambiguously assigned to the ^{1}H signal at 5.82 ppm and so we write for site 24: $\delta_H = 5.82$ ppm and $\delta_C = 117.5$ ppm.

The site 21 methylene group is alpha to an oxygen and, as such, generates the most downfield methylene signals in the molecule. We predict the ^{13}C chemical shift of this methylene group will be greater than 60 ppm and so we write for site 21: $\delta_H = 4.83$ & 4.92 ppm and $\delta_C = 74.1$ ppm.

Normally we are able to immediately identify carbonyl ^{13}C resonances, but for this molecule we find two ^{13}C resonances as essentially equal probability candidates. We observe two nonprotonated ^{13}C signals at 174.6 and 176.8 ppm which we attribute to sites 20 and 23 in no particular order. We can again use the argument that C20, being fewer bonds from the ^{1}H's on ring IV (H17, H16's, etc.), should correspondingly generate a signal that participates in a greater number of gHMBC cross peaks than does C23. If we examine the 2-D ^{1}H-^{13}C gHMBC NMR spectrum of digoxigenin, we observe that the ^{13}C signal at 176.8 ppm participates in

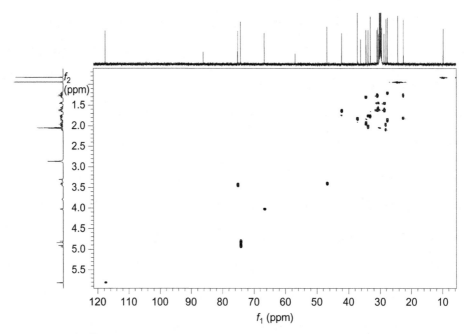

FIG. 2.9.7 The 2-D ^{1}H-^{13}C HSQC NMR spectrum of digoxigenin in acetone-d_6.

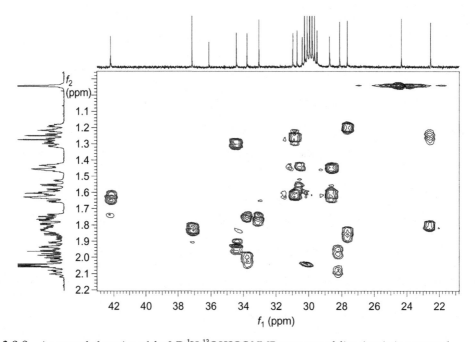

FIG. 2.9.8 An expanded portion of the 2-D ^{1}H-^{13}C HSQC NMR spectrum of digoxigenin in acetone-d_6.

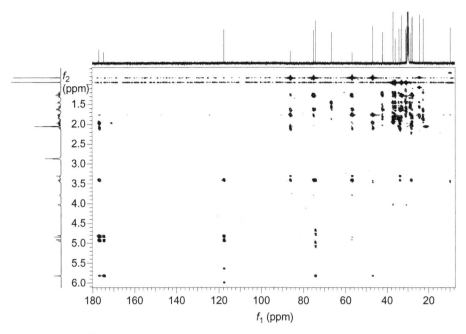

FIG. 2.9.9   The 2-D ^{1}H-^{13}C gHMBC NMR spectrum of digoxigenin in acetone-d_6.

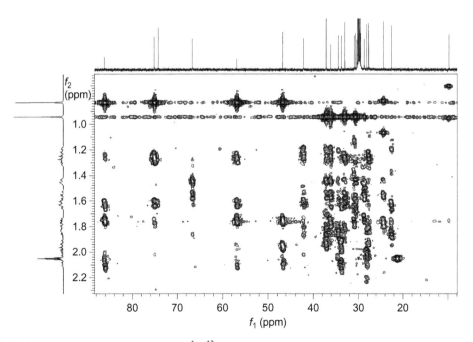

FIG. 2.9.10   An expanded portion of the 2-D ^{1}H-^{13}C gHMBC NMR spectrum of digoxigenin in acetone-d_6.

TABLE 2.9.1   Predicted Multiplicities for the ^{1}H's of Digoxigenin

Site in Molecule	Expected Multiplicity
1	td, dt
2	qd, dq
3	tt (OH s)
4	td, d^3
5	tt
6	qd, dq
7	qd, dq
8	td
9	td
11	td, dt
12	d^2 (OH s)
14	(OH s)
15	td, dt
16	qd, dq
17	d^2
18	s
19	s
21	2×d
24	s

TABLE 2.9.2   Carbons of Digoxigenin

Type of Carbon	Site Number
CH_3 (methyl)	18, 19
CH_2 (methylene)	1, 2, 4, 6, 7, 11, 15, 16, 21
CH (methine)	3, 5, 8, 9, 12, 17, 24
C_{np} (nonprotonated)	10, 13, 14, 20, 23

seven cross peaks while the ^{13}C signal at 174.6 ppm participates in only four cross peaks. Based on the more social nature of the ^{13}C signal at 176.8 ppm, we assign this signal to C20 and the signal at 174.6 ppm to C23. We therefore write for site 20: $\delta_C = 176.8$ ppm and for site 23: $\delta_C = 174.6$ ppm.

One of the cross peaks involving the C20 signal (at 176.8 ppm) in the gHMBC spectrum also involves a methine ^{1}H signal at 3.40 ppm with a ^{13}C signal with a chemical shift of 46.7 ppm.

TABLE 2.9.3 ^{1}H and ^{13}C NMR Signals of Digoxigenin Listed by Group Type

^{1}H Signal (ppm)	^{13}C Signal (ppm)	Group Type
–	176.8	Nonprotonated
–	174.6	Nonprotonated
5.82	117.5	Methine
–	86.1	Nonprotonated
3.44	75.0	Methine
4.83 & 4.92	74.1	Methylene
4.04	66.6	Methine
–	56.8	Nonprotonated
3.40	46.7	Methine
1.63	42.2	Methine
1.82	37.1	Methine
–	36.1	Nonprotonated
1.30 & 1.93	34.4	Methylene
1.75 & 2.01	33.7	Methylene
1.76	33.0	Methine
1.26 & 1.61	30.9	Methylene
1.44 & 1.57	30.7	Methylene
1.45 & 1.62	28.7	Methylene
1.97 & 2.09	28.1	Methylene
1.20 & 1.86	27.6	Methylene
0.94	24.3	Methyl
1.26 & 1.81	22.5	Methylene
0.83	9.8	Methyl

Given the choice between H8 which is five bonds distant, H12 which is four bonds distant and has a ^{13}C shift that is too low, and H17 that is only two bonds away, we assign this methylene signal to site 17: $\delta_H = 3.40$ ppm and $\delta_C = 46.7$ ppm. The signals of the H21's only correlate with the signals of C20, C23, and C24 in the gHMBC spectrum so we cannot confirm our site 17 assignment with one or both gHMBC cross peaks between the signals of the H21's and that of C17. The H24 signal at 5.82 ppm shares a gHMBC cross peak with the C17 signal at 46.7 ppm.

We are able to identify the H18 methyl ^{1}H signal using the gHMBC spectrum. The more upfield of the two methyl ^{1}H signals at 0.83 ppm shares a strong gHMBC correlation with the C17 signal at 46.7 ppm. We assign this methyl ^{1}H signal to site 18: $\delta_H = 0.83$ ppm and $\delta_C = 9.8$ ppm.

The H18 signal at 0.83 ppm also shares prominent gHMBC cross peaks with nonprotonated ^{13}C signals at 86.1 and 56.8 ppm. We have three nonprotonated carbon sites yet to be identified, and so it seems reasonable to assume that these two ^{13}C signals are those of sites 13 and 14, because C13 and C14 are two and three bonds separated from the H18's, while the C10 is six bonds removed from the H18's. Site 14 bears an oxygen atom while site 13 does not, and so we assign the more downfield ^{13}C shift to C14 because of the electron density withdrawal caused by the site 14 oxygen. We write for site 14: $\delta_C = 86.1$ ppm and for site 13: $\delta_C = 56.8$ ppm. A partial confirmation is obtained by noting that the gHMBC cross peak from the H17 signal at 3.40 ppm to the C13 signal at 56.8 ppm is strong compared with that between the signals of H17 and C14 (86.1 ppm).

By process of elimination we are now able to assign the sole unassigned nonprotonated ^{13}C signal to site 10: $\delta_C = 36.1$ ppm.

In the gHMBC spectrum, the methine ^{1}H resonance at 3.44 ppm shares a cross peak with the C17 signal at 46.7 ppm and also with the C13 signal at 9.8 ppm. The ^{13}C shift of the methine group with the ^{1}H signal at 3.44 ppm is 75.0 ppm, suggesting that an oxygen is attached to this methine group. Given the choice between assigning these shifts to site 12 and site 8, we clearly choose site 12 because site 8 has no oxygen atom alpha to it. We write for site 12: $\delta_H = 3.44$ ppm and $\delta_C = 75.0$ ppm. Examination of the gCOSY spectrum reveals a cross peak betweem a hydroxyl ^{1}H signal at 3.79 ppm and the methine ^{1}H signal at 3.44. We assign the 3.79 ppm signal to the hydroxyl group of site 12.

In the midfield region of the HSQC spectrum we only have one other cross peak from the signal of a ^{1}H on a carbon with an alpha oxygen atom. By elimination we assign the methine signal at 4.04 ppm to site 3: $\delta_H = 4.04$ ppm and $\delta_C = 66.6$ ppm.

Also by the process of elimination we are able to assign our other methyl group at site 19. We write for site 19: $\delta_H = 0.94$ ppm and $\delta_C = 24.3$ ppm. We confirm this assignment by noting that the H19 signals at 0.94 ppm share a gHMBC cross peak with the signals of the methyl attachment point of C10 at 35.6 ppm.

We are now faced with the challenge of assigning the last three methine groups at site 5, 8, and 9. These three ^{1}H/^{13}C shift pairs are found at the following shift values: 1.63/42.2, 1.76/33.0, and 1.82/37.1 ppm. The H19 signals are expected to share gHMBC cross peaks with the signals of C5 and C9, because C5 and C9 are both three bonds away from the H19's, while C8 is four bonds distant. The H19 signals share gHMBC cross peaks with the ^{13}C methine signals at 33.0 and 37.1 ppm, so these must be the signals of C5 and C9 in no particular order. By elimination we conclude that for site 8: $\delta_H = 1.63$ ppm and $\delta_C = 42.2$ ppm. We observe gHMBC data that is not inconsistent with our assignment of site 8. The ^{1}H resonance at 1.63 ppm shares a strong gHMBC cross peak with the nonprotonated C14 resonance at 86.1 ppm, but so does a ^{1}H signal at 1.76 ppm (there are two ^{1}H signals at this chemical shift).

The methine ^{1}H resonance at 1.76 ppm generates gHMBC cross peaks with the signals of C12 (75.0 ppm) and C19 (24.3 ppm). This must therefore be the ^{1}H signal of H9, because H9 is three bonds distant from both C12 and C19. We write for site 9: $\delta_H = 1.76$ ppm and $\delta_C = 33.0$ ppm.

By elimination we now write for site 5: $\delta_H = 1.82$ ppm and $\delta_C = 37.1$ ppm. There is a disappointing lack of unambiguous gHMBC cross peaks that might confirm this assignment. If the junction between rings I and II were *trans*, we would expect to see a significant gHMBC cross peak between the signals of H5 and C19.

The H9 methine resonance at 1.76 ppm shares a significant cross peak with the methylene ^{13}C resonance at 30.9 ppm. We assign this methylene group as that of site 11: $\delta_H = 1.26$ & 1.61 ppm and $\delta_C = 30.9$ ppm. The H11 signals at 1.26 & 1.61 ppm share confirming gHMBC cross peaks with the signals of C12 (75.0 ppm), C13 (56.8 ppm), and C9 (33.0 ppm).

The H17 methine ^{1}H signal shares gHMBC cross peaks with two methylene ^{13}C signals at 28.1 and 33.7 ppm. The 28.1 ppm ^{13}C signal is assigned to site 16 and the 33.7 ppm signal to site 15. Both of the ^{1}H's (1.97 & 2.09 ppm) associated with the 28.1 ppm ^{13}C signal (in the HSQC) correlate in the gHMBC spectrum with the C17 signal at 46.7 ppm and so we write for site 16: $\delta_H = 1.97$ & 2.09 ppm and $\delta_C = 28.1$ ppm and for site 15: $\delta_H = 1.75$ & 2.01 ppm and $\delta_C = 33.7$ ppm. The H15's do not generate signals that participate in particularly useful or convincing COSY or gHMBC cross peaks to allow confirmation of the site 15 versus site 16 assignment. It is noted, however, that chemical prediction software confirms that the signal of C15 should be downfield from that of C16 by six ppm, as we observe experimentally.

The methine H8 signal shares a gHMBC cross peak with a methylene ^{13}C signal at 22.5 ppm. We attribute this signal to C7 and write for site 7: $\delta_H = 1.26$ & 1.81 ppm and $\delta_C = 22.5$ ppm.

The well-resolved methylene ^{1}H signal at 1.20 ppm shares gHMBC cross peaks with the signals of C8 (42.2 ppm), C5 (37.1 ppm), and C10 (36.1 ppm). We assign the ^{1}H signal at 1.20 ppm to site 6: $\delta_H = 1.20$ & 1.86 ppm and $\delta_C = 27.6$ ppm.

We now have three methylene groups left to assign at sites 1, 2, and 4. Because we have already identified the H3 signal as being that at 4.04 ppm, we are able to use the COSY spectrum to see that the H3 signal correlates with four ^{1}H signals with chemical shifts of 1.31, 1.45, 1.63, and 1.94 ppm. By process of elimination, we are able to uniquely identify the H1 signals as being those unassigned methylene ^{1}H signals that do not couple with the H3 signal at 4.04 ppm. We write for site 1: $\delta_H = 1.44$ & 1.57 ppm and $\delta_C = 30.7$ ppm.

The methylene ^{1}H signal at 1.93 ppm is observed to couple in the gHMBC spectrum to the C5 signal at 37.1 ppm. This allows us to assign the ^{1}H shift at 1.93 ppm to one of H4's and write for site 4: $\delta_H = 1.30$ & 1.93 ppm and $\delta_C = 34.4$ ppm. By elimination, we write for site 2: $\delta_H = 1.45$ & 1.62 ppm and $\delta_C = 28.7$ ppm. The H5 signal at 1.82 ppm likely shares a gHMBC cross peak with the C4 resonance at 34.4 ppm, for although the H7 signal at 1.81 ppm may correlate with the C15 signal at 33.7 ppm, we note that H7 the precision by which we can measure gHMBC cross peak position and also H7 is four bonds away from C15, suggesting that this moderately intense gHMBC cross peak cannot plausibly be ascribed to the H7-C16 interaction. We also observe gHMBC correlations between the H2 signals at 1.45 & 1.62 ppm and the C5 signal at 37.1 ppm.

3

Alkaloids, Nonsteroidal Hormones, Drugs, and Glycosides

Corticosteroids and polypeptides cooperate with human attempts at categorization more readily than do the molecules appearing in this chapter. This chapter contains the spectra of large molecules that fail to fall neatly into other categories.

3.1 ISORESERPINE IN ACETONE-D_6

Isoreserpine is notable for its relatively large size. Its biosynthesis involves tryptophan, and it presumably exerts physiological effects. The structure of isoreserpine is shown in Fig. 3.1.1. With six rings—three of which are aromatic, with two forming an indole heterocycle, plus six methoxy groups, two ester linkages, and two nitrogens, this $C_{33}H_{40}N_2O_9$ compound appears initially intimidating but, with the application of our systematic assignment methodology, yields its shift-site pairings.

A small amount of isoreserpine was dissolved in acetone-d_6 and used to generate all of the NMR spectra in this section. The 1-D ^1H NMR spectrum is shown in Fig. 3.1.2. The 1-D ^{13}C NMR spectrum appears in Fig. 3.1.3. The 2-D ^1H-^1H COSY NMR spectrum of isoreserpine is displayed in Fig. 3.1.4. The 2-D ^1H-^{13}C HSQC NMR spectrum is shown in Fig. 3.1.5, and finally we have the 2-D ^1H-^{13}C gHMBC NMR spectrum of isoreserpine in Fig. 3.1.6.

We generate our semiobligatory three tables of data that we derive from the information contained in the NMR spectra of isoreserpine. Table 3.1.1 presents our predictions for the multiplicities of the ^1H resonances. Table 3.1.2 presents a listing of group type by numbered molecular site. Using information in the 1-D ^1H and ^{13}C NMR spectra and also in the 2-D ^1H-^{13}C HSQC NMR spectrum, we list all ^{13}C chemical shifts from the solute, matching ^{13}C signals with those of their attached ^1H's based on the cross peaks observed in the HSQC spectrum (Table 3.1.3).

Initial inspection of the 1-D ^1H NMR spectrum of isoreserpine in Fig. 3.1.1 shows rather poor integral calibration, with values that both exceed and fall well short of expectation (integer) values. The diminution of this valuable integral accounting tool is a setback but not insurmountable. The joke is that adversity enhances personal character, just as pain begets art.

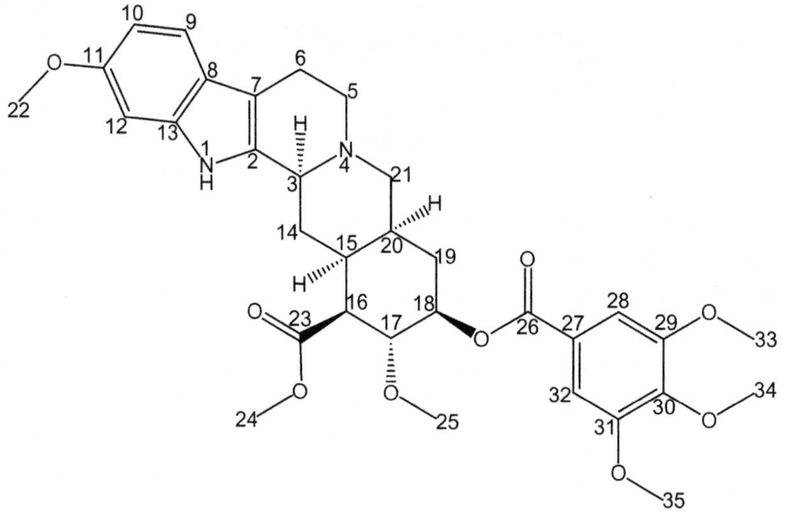

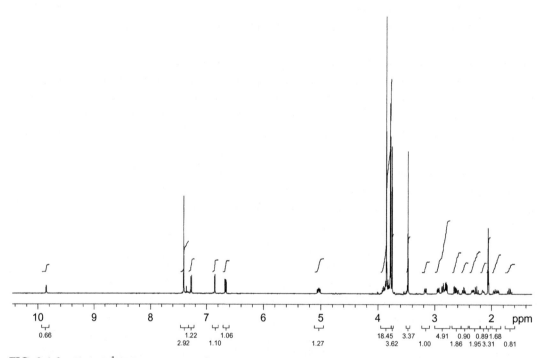

FIG. 3.1.1   The structure of isoreserpine.

FIG. 3.1.2   The 1-D ^{1}H NMR spectrum of isoreserpine in acetone-d_6.

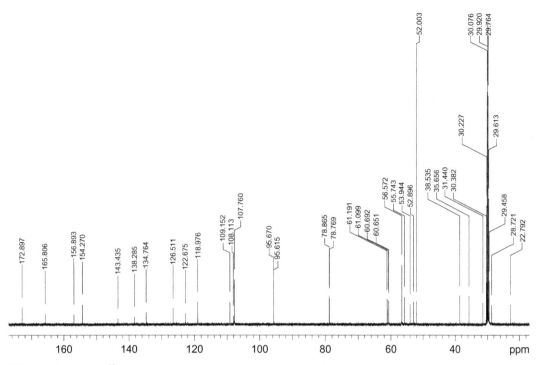

FIG. 3.1.3 The 1-D ^{13}C NMR spectrum of isoreserpine in acetone-d_6.

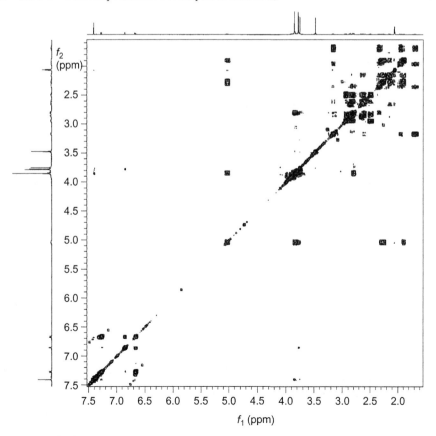

FIG. 3.1.4 The 2-D ^1H-^1H COSY NMR spectrum of isoreserpine in acetone-d_6.

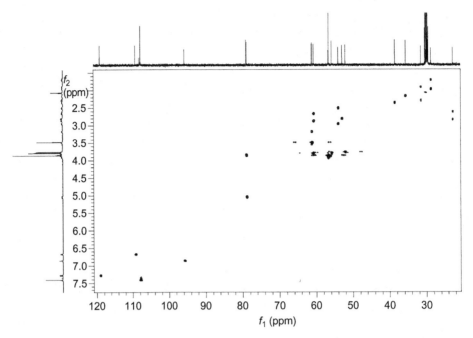

FIG. 3.1.5   The 2-D ^{1}H-^{13}C HSQC NMR spectrum of isoreserpine in acetone-d_6.

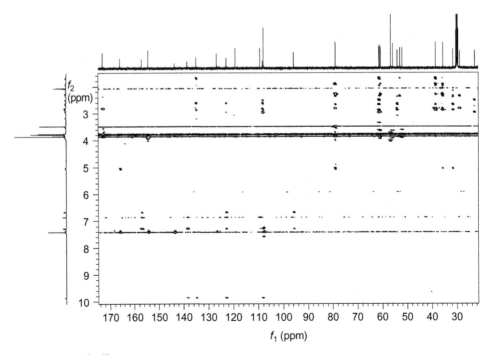

FIG. 3.1.6   The 2-D ^{1}H-^{13}C gHMBC NMR spectrum of isoreserpine in acetone-d_6.

TABLE 3.1.1   Predicted Multiplicities for the ^1H Resonances of Isoreserpine

Site in Molecule	Expected Multiplicity
1	s
3	d^2
5	td, dt
6	td, dt
9	d
10	d
12	s
14	$2 \times d^3$
15	d^4
16	d^2
17	d^2
18	d^3
19	$2 \times d^3$
20	d^5
21	$2 \times d^2$
22	s
24	s
25	s
28/32	s
33/35	s
34	s

TABLE 3.1.2   Numbered Sites of Isoreserpine

Type of Carbon	Site Number
CH_3 (methyl)	22, 24, 25, 33, 34, 35
CH_2 (methylene)	5, 6, 14, 19, 21
CH (methine)	3, 9, 10, 12, 15, 16, 17, 18, 20, 28, 32
C_{np} (nonprotonated)	2, 7, 8, 11, 13, 23, 26, 27, 29, 30, 31
Numbered Heteronuclei	1, 4

TABLE 3.1.3   ^{1}H and ^{13}C NMR Signals of Isoreserpine Listed by Group Type

[1]H Signal (ppm)	[13]C Signal (ppm)	Group Type
–	172.9	Nonprotonated
–	165.8	Nonprotonated
–	156.9	Nonprotonated
–	154.3[a]	Nonprotonated
–	143.4	Nonprotonated
–	138.3	Nonprotonated
–	134.8	Nonprotonated
–	126.5	Nonprotonated
–	122.7	Nonprotonated
7.28	119.0	Methine
6.68	109.2	Methine
–	108.1	Nonprotonated
7.41	107.8[a]	Methine
6.87	95.6[b]	Methine
5.05	78.9	Methine
3.85	78.8	Methine
3.48	61.2	Methyl
3.17	61.1	Methine
3.78	60.69	Methyl
2.68 & 2.89	60.65	Methylene
3.84	56.6[a]	Methyl
3.77	55.7	Methyl
2.49 & 2.97	53.9	Methylene
2.81	52.9	Methine
3.75	52.0	Methyl
2.33	38.5	Methine
2.16	35.7	Methine
1.92 & 2.27	31.4	Methylene
1.69 & 1.95	28.7	Methylene
2.60 & 2.83	22.8	Methylene

[a]Doubly intense.
[b]Determined to be a signal from just one ^{13}C site.

The higher educational experience is a testament to this credo, and for some it becomes a way of life. Choose carefully. For now, we can just focus on this problem.

One low integral ^{1}H resonance we notice immediately is the downfield signal at 9.86 ppm with an integral of 0.66. We assign this signal to the site 1 indole proton. The integrals of ^{1}H's on heteroatoms are often low no matter how carefully we set our NMR acquisition parameters (to guard against nonlinear receiver response, short relaxation delay, etc.). We can explain the chemical shift of the ^{1}H on nitrogen as a result of the combination of an aromatic ring current shift and the electronegativity of the nitrogen atom—this is a unique chemical environment in this molecule. We write for site 1: $\delta_H = 9.86$ ppm.

A second useful entry point for this 33-carbon behemoth involves first recognizing the c_2 rotational symmetry axis for the aromatic ring at the lower right portion molecule as it appears in Fig. 3.1.1. The homotopic relationship between sites 28 and 32, between 29 and 31, and between 33 and 35, generates three sets of double intensity NMR signals that are easy to identify. The downfield ^{1}H singlet at 7.41 ppm that integrates to 2.92 (the integral includes a nearby impurity) is attributed to the signal of the H28/32 sites. We write for sites 28/32: $\delta_H = 7.41$ and $\delta_C = 107.8$ ppm.

We also can locate the double intensity methoxy ^{1}H signal at 3.84 ppm. We assign the NMR signals for sites 33/35 as follows: $\delta_H = 3.84$ ppm and $\delta_C = 56.6$ ppm.

Our last double intensity signals are for sites 29/31 and are found in the 1-D ^{13}C NMR spectrum at 154.3 ppm. We assign sites 29/31 to the signal $\delta_C = 154.3$ ppm.

Having started with five, we still have three unassigned methoxy ^{1}H signals very close to one another on the chemical shift axis at 3.75, 3.77, and 3.78 ppm. The methoxy ^{1}H signal at 3.78 ppm shares a gHMBC correlation with the C29/31 ^{13}C signal at 154.3 ppm. This ^{1}H methoxy signal must therefore be that of site 34, because no other methoxy groups are nearby. We write for site 34: $\delta_H = 3.78$ ppm and $\delta_C = 60.69$ ppm.

The site 34 methoxy group attaches to the lower-right aromatic ring of isoreserpine at site 30. The H34 signal shares a gHMBC cross peak with a nonprotonated ^{13}C signal at 143.4 ppm. We assign this signal to site 30: $\delta_C = 143.4$ ppm.

Having assigned the signals of the lower-right ring of isoreserpine as it is shown in Fig. 3.1.1, we are able to use the signals of the aromatic ring ^{1}H's to differentiate between the two carbonyl ^{13}C signals. In the gHMBC spectrum, the very sharp ^{1}H signal from H28 and H32 at 7.41 ppm shares a strong gHMBC correlation with the nonprotonated ^{13}C signal at 165.8 ppm. We assign this ^{13}C signal to the site 26 carbonyl: $\delta_C = 165.8$ ppm.

The other most downfield ^{13}C signal must arise from site 23 and so we write for site 23: $\delta_C = 172.9$ ppm.

Having identified C23, we are able to locate the H24 methoxy ^{1}H signal that correlates with C23 in the gHMBC spectrum. We observe a gHMBC cross peak between a methoxy ^{1}H signal at 3.75 ppm and the C23 signal at 172.9 ppm. We assign site 24 as follows: $\delta_H = 3.75$ ppm and $\delta_C = 52.0$ ppm.

The methine group at site 16 is alpha to the C23 carbonyl and so the H16 and C23 signals are expected to correlate strongly in the gHMBC spectrum. We observe a gHMBC cross peak between the methine ^{1}H signal at 2.81 ppm and the C23 signal at 172.9 ppm, allowing us to write for site 16: $\delta_H = 2.81$ ppm and $\delta_C = 52.9$ ppm.

We have thus far assigned all but two of the methoxy groups. If we compare the attachment points for the site 22 versus the site 25 methoxy groups, we note that C11 is an aromatic carbon and C17 is aliphatic. The methoxy ^{1}H signal at 3.48 ppm shares a gHMBC cross peak with a ^{13}C signal at 78.8 ppm—too far upfield to be considered an aromatic ^{13}C chemical shift. We assign the ^{1}H shift to site 25 and the ^{13}C shift to site 17, allowing us to write for site 25: $\delta_H = 3.48$ ppm and $\delta_C = 61.2$ ppm and for site 17: $\delta_H = 3.85$ ppm and $\delta_C = 78.8$ ppm.

The top-left portion of isoreserpine has ^{1}H's arranged on the aromatic ring in a 1,2,4 substitution pattern. The middle member (in terms of location in the molecule, not along the chemical shift axis) of this three-member spin system (in this case H10) generates a signal that appears as a doublet of doublets. If we examine the COSY spectrum we can readily locate the ^{1}H NMR signals from these three members of this spin system at chemical shifts of 7.28, 6.87, and 6.68 ppm. Examination of the multiplet splitting patterns in the 1-D ^{1}H NMR spectrum reveals that the ^{1}H signal at 6.68 ppm is split into a doublet of doublets. We assign this signal to H10 because it has a strong cis-$^3J_{HH}$ to H9 and a weaker $^4J_{HH}$ to H12. We write for site 10: $\delta_H = 6.68$ ppm and $\delta_C = 109.2$ ppm.

The more strongly split doublet in the H9/H10/H12 spin system must be H9 signal. The ^{1}H signal at 7.28 ppm is clearly split more than the signal at 6.87 ppm and so we can write for site 9: $\delta_H = 7.28$ ppm and $\delta_C = 119.0$ ppm.

The last spin in this three-member spin system (we neglect the dilute ^{13}C's when we discuss ^{1}H multiplet appearances in the context of spin systems) is H12, so for site 12: $\delta_H = 6.87$ ppm and $\delta_C = 95.6$ ppm.

The H9 should couple, through a trans-$^3J_{HC}$, to C11, as should the H22's. We observe that the H9 signal at 7.28 ppm shares a gHMBC cross peak with a nonprotonated ^{13}C signal at 156.9 ppm. Also observable in the gHMBC spectrum is a correlation (the t_1 ridge in the plot makes it hard to see) from the ^{1}H signal at 3.77 ppm to the same nonprotonated ^{13}C signal at 156.9 ppm. The H10 and H12 signals also correlate with this nonprotonated ^{13}C signal, sharing cross peaks just above the cross peak between the H9 and C11 signals in the gHMBC spectrum. We write for site 11: $\delta_C = 156.9$ ppm.

We also have now determined the site 22 assignments: $\delta_H = 3.77$ ppm and $\delta_C = 55.7$ ppm.

Of the four remaining methines (sites 3, 15, 18, and 20), sites 15 and 20 should generate the most upfield signals and the site 18 signals be more downfield compared with the site 3 signals. That is, the site 18 methine is bound to an oxygen and so its signal should be furthest downfield, while the site 3 methine is bound to a not-as-electronegative nitrogen atom. The site 15 and 20 methines are bracketed by sp^3-hybridized carbons and so are expected to generate ^{1}H signals which are more upfield than the signals of the methines alpha to electronegative heteroatoms. In the midfield region of the HSQC spectrum we observe two methine cross peaks with ^{13}C chemical shifts near 78 ppm. The more downfield ^{1}H signal at 5.05 ppm we attribute to site 18, writing $\delta_H = 5.05$ ppm and $\delta_C = 78.9$ ppm. We are able to confirm this assignment by noting that the ^{1}H signal at 5.05 ppm shares a gHMBC cross peak with the C26 signal at 165.8 ppm.

In the COSY spectrum, the H18 signal at 5.05 ppm correlates with two ^{1}H signals at 1.92 and 2.27 ppm which are seen to be shifts from a single methylene group. We take these methylene ^{1}H signals to be those of site 19 and write: $\delta_H = 1.92$ & 2.27 ppm and $\delta_C = 31.4$ ppm.

Continuing on with our assignment of the remaining methine groups, we select the only other midfield HSQC methine signal and assign this to site 3: $\delta_H = 3.17$ ppm and $\delta_C = 61.1$ ppm.

A pair of upfield ^{1}H signals from a methylene group is observed, in the COSY spectrum, to correlate with the H3 signal at 3.17 ppm. We assign these methylene ^{1}H signals to the H14's and write for site 14: $\delta_H = 1.69$ & 1.95 ppm and $\delta_C = 28.7$ ppm.

In the COSY spectrum, we see a methine ^{1}H signal at 2.33 ppm that shares cross peaks with both H14 signals at 1.69 & 1.95 ppm. We assign this methine ^{1}H signal to site 15: $\delta_H = 2.33$ ppm and $\delta_C = 38.5$ ppm.

Our last methine group is at site 20. We can confirm that our assignment of the shifts of sites 3, 14, 15, and 20 is at least self-consistent by noting that in the COSY spectrum, the last unassigned methine ^{1}H signal at 2.16 ppm correlates with the H19 signals at 1.92 & 2.27 ppm. We write for site 20: $\delta_H = 2.16$ ppm and $\delta_C = 35.7$ ppm.

We have five nonprotonated ^{13}C signals to pair with sites 2, 7, 8, 13, and 27. The ^{1}H at site 1, residing on the nitrogen, does not exchange so quickly in the aprotic solvent (acetone-d_6) as to prevent us from using the H1 signal in the gHMBC spectrum to identify the ^{13}C signals of sites 2, 7, 8, and 13. We observe correlations from the H1 signal at 9.86 ppm to signals from nonprotonated ^{13}C's at 138.3, 134.8, 122.7, and 108.1 ppm. By process of elimination, we recognize that the C27 signal must be from the ^{13}C signal at 126.5, because this is the sole unassigned ^{13}C signal that does not correlate with the H1 signal. We write for site 27: $\delta_C = 126.5$ ppm. We are able to confirm our C27 assignment by noting the presence of a gHMBC cross peak between the very sharp and intense H28/32 signals and the C27 signal. C27 is two bonds removed from the H28/32's but, because C28 & C32 are sp^2-hybridized, C7 and the H28/32's are expected to couple only weakly due to the low value for the $^2J_{CH}$ because of the 120 degrees bond angle from H28 to C27 and from H32 to C27.

C2 and C13 are anticipated to generate ^{13}C signals further downfield compared with those of C7 and C8. The H9 signal at 7.28 ppm shares two very strong gHMBC correlations, one of which is with the most downfield and still-unassigned nonprotonated ^{13}C signals. The intense gHMBC correlation between the H9 and C11 is expected because a large $trans$-$^3J_{CH}$ exists between H9 and C11. A similar coupling is expected between H9 and C13. We observe a second strong gHMBC cross peak between the H9 signal at 7.28 ppm and a ^{13}C signal at 138.3 ppm. We assign this signal to site 13: $\delta_C = 138.3$ ppm.

One of the two H14 signals (at 1.69 ppm) shares a gHMBC cross peak with the next most downfield, still-unassigned nonprotonated ^{13}C signal at 134.8 ppm (our other two still-unassigned ^{13}C signals are found at 108.1 and 122.7 ppm). We assign the 134.8 ppm ^{13}C signal to site 2: $\delta_C = 134.8$ ppm.

We observe another strong gHMBC cross peak between the H10 signal at 6.68 ppm and the ^{13}C signal at 122.7 ppm. Because the geometric relationship between H10 and C8 is expected to generate a large $trans$-$^3J_{CH}$ similar to that between H9 and C11 and also between H9 and C13, we assign the ^{13}C signal at 122.7 ppm to site 8: $\delta_C = 122.7$ ppm. The H10 signal also shares a comparable gHMBC cross peak with the C12 signal, as we expect based on our understanding of the magnitude of the coupling interaction when a $trans$-$^3J_{CH}$ exists.

We assign the last remaining nonprotonated ^{13}C signal at 108.1 ppm to that of site 7: $\delta_C = 108.1$ ppm. We are able to confirm this assignment by noting that the slightly less intense cis-$^3J_{CH}$ from H9 to C7 generates a significant gHMBC cross peak between the H9 signal at 7.28 ppm and the ^{13}C signal of C7 at 108.1 ppm.

There are but three remaining methylene groups to assign. Moving from left to right in the HSQC spectrum (f_1 is horizontal), we see the first set of methylene signals at $\delta_H = 2.68$ & 2.89 ppm and $\delta_C = 60.65$ ppm, the second set of signals at $\delta_H = 2.49$ & 2.97 ppm and $\delta_C = 53.9$ ppm, and the third set at $\delta_H = 2.60$ & 2.83 ppm and $\delta_C = 22.8$ ppm. These three sets of methylene NMR signals must be paired with sites 5, 6, and 21 (with no implied order).

Because the methylene groups at sites 5 and 21 are both bound to the nitrogen atom at site 1, the chemical shifts of their signals are expected to be similar and downfield from the site 6 shifts. While all three sets of ^{1}H shifts are similar for these last three methylene groups, one of the ^{13}C shifts is found markedly upfield at 22.8 ppm. We assign site 6 based on this upfield ^{13}C shift as follows: $\delta_H = 2.60$ & 2.83 ppm and $\delta_C = 22.8$ ppm.

In the gHMBC spectrum, the C6 resonance at 22.8 ppm shares cross peaks with the signals of only two ^{1}H's. The ^{1}H signals that correlate with the C6 signal have chemical shifts of 2.49 & 2.97 ppm. These ^{1}H shifts are assigned to site 5: $\delta_H = 2.49$ & 2.97 ppm and $\delta_C = 53.9$ ppm. The lack of gHMBC cross peaks with the C6 signal as participant makes sense if we examine the structure of isoreserpine and note how far removed site 6 is from any ^{1}H's except the H5's.

By process of elimination, we assign the last set of methylene NMR signals to site 21: $\delta_H = 2.68$ & 2.89 ppm and $\delta_C = 60.65$ ppm. Comparison of the chemical environment of site 21 versus site 5 reveals that the NMR signals from spins at site 21 should be slightly downfield (at greater ppm values) compared with those spins from site 5. Alas, the gHMBC cross peak between the signals of H20 and C21 is not observed, perhaps due to steric crowding and bond angle distortion. The methine ^{1}H signal from site 15 observed at 2.33 ppm fails to correlate in the gHMBC spectrum with the C21 signal at 60.65 ppm. However, the H19 signals at 1.92 & 2.27 ppm share gHMBC cross peaks with the C21 signal at 60.65 ppm. The more downfield of the two H5 methylene ^{1}H signals at 2.97 ppm, believed to be that of the equatorial and not the axial H5, shares a gHMBC cross peak with the C21 signal at 60.65 ppm, thus confirming the assignments of sites 5, 6, and 21.

3.2 LOVASTATIN IN CHLOROFORM-D

Lovastatin is man-made. It is a 24-carbon molecule with three rings, two ester function-alities with one being a lactone, a pair of conjugated carbon-carbon double bonds, and a hydroxyl group. With an IHD of seven, the empirical formula of lovastatin is $C_{24}H_{36}O_5$. The structure of lovastatin is shown in Fig. 3.2.1. A small amount of lovastatin was dissolved in chloroform-d to make the sample. The 1-D ^{1}H NMR spectrum of this sample is shown in Fig. 3.2.2, with a second rendering more amenable to multiplet inspection in Fig. 3.2.3. The 1-D ^{13}C NMR spectrum of the sample is shown in Fig. 3.2.4. The 2-D ^{1}H-^{1}H COSY NMR spectrum of the sample is shown in Fig. 3.2.5, with two additional versions containing expanded portions of the spectrum in Figs. 3.2.6 and 3.2.7. The 2-D ^{1}H-^{13}C HSQC NMR spectrum of the sample is shown in Fig. 3.2.8. An expanded portion of the same HSQC spectrum appears in Fig. 3.2.9. The 2-D ^{1}H-^{13}C gHMBC NMR spectrum is shown in Fig. 3.2.10. An expanded version of the gHMBC spectrum is found in Fig. 3.2.11.

To begin, we examine the structure of lovastatin in Fig. 3.2.1 and place our predictions of the ^{1}H resonance multiplicities into Table 3.2.1. Based on the structure of lovastatin in Fig. 3.2.1, we also list an accounting of the numbered molecular sites of the molecule in Table 3.2.2. Table 3.2.3 contains the pairing of the ^{1}H and ^{13}C signals afforded by information in the HSQC spectrum supplemented with information on nonprotonated carbon sites from the 1-D ^{13}C spectrum.

There are two entry points we can identify: (1) the unique splitting of the H14 signal by the H13's to generate the only methyl ^{1}H signal split into a triplet, and (2) the conjugated carbon-carbon double bonds.

We begin with the triplet methyl ^{1}H signal. Note that the H14's generate the only methyl ^{1}H signal predicted to be a doublet of doublets (or a triplet). The most upfield methyl ^{1}H signal at 0.88 ppm is a triplet and so we write for site 14: $\delta_H = 0.88$ ppm and $\delta_C = 11.9$ ppm.

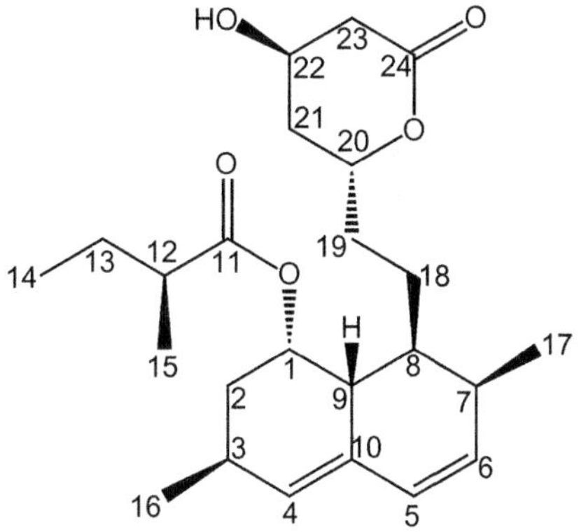

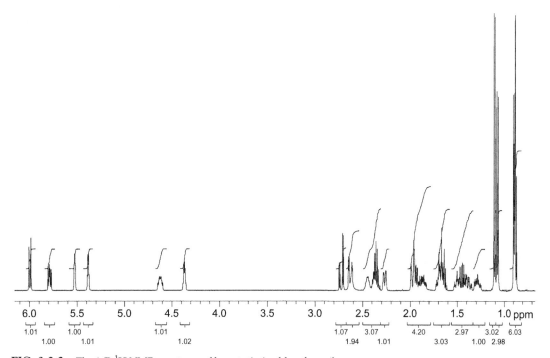

FIG. 3.2.1 The structure of lovastatin.

FIG. 3.2.2 The 1-D ^{1}H NMR spectrum of lovastatin in chloroform-d.

The COSY spectrum contains a cross peak between the H14 signal at 0.88 ppm and methylene ^{1}H signals at 1.44 & 1.66 ppm. We assign these signals to site 13: $\delta_H = 1.44$ & 1.66 ppm and $\delta_C = 27.0$ ppm.

The H13 signals at 1.44 & 1.66 ppm share gHMBC cross peaks with the ^{13}C signal from a methine group at 41.7 ppm. We assign this ^{13}C signal to C12 and write for site 12: $\delta_H = 2.35$ ppm and $\delta_C = 41.7$ ppm.

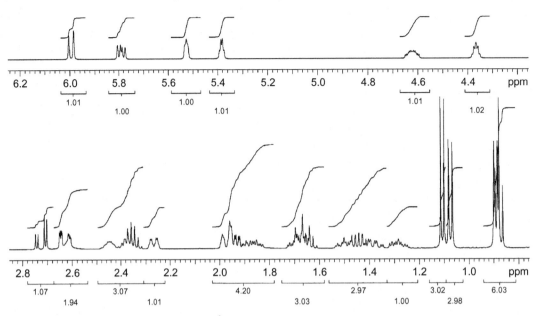

FIG. 3.2.3　An expanded portion of the 1-D ^{1}H NMR spectrum of lovastatin in chloroform-d.

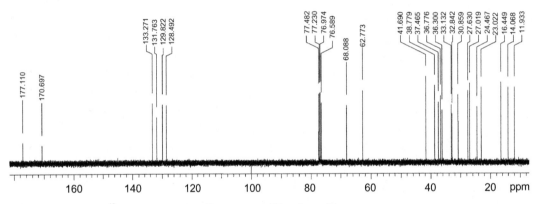

FIG. 3.2.4　The 1-D ^{13}C NMR spectrum of lovastatin in chloroform-d.

The methyl ^{1}H signal at 1.11 ppm also shares a gHMBC cross peak with the C12 signal at 41.7 ppm. This methyl ^{1}H signal is attributed to site 15: $\delta_H = 1.11$ ppm and $\delta_C = 16.4$ ppm.

The H15 resonance at 1.11 ppm shares a gHMBC cross peak with the most downfield ^{13}C resonance at 177.1 ppm. We assign this signal to the carbonyl at site 11: $\delta_C = 177.1$ ppm.

By process of elimination, we assign the other carbonyl carbon (the lactone) at site 24: $\delta_C = 170.7$ ppm. Because site 10 is the only other nonprotonated carbon site to assign, we write for site 10: $\delta_C = 131.8$ ppm.

The ^{1}H signals from a methylene group at 2.63 & 2.72 ppm participate in a pair of strong cross peaks with the C24 signal at 170.7 ppm in the gHMBC spectrum. We assign these methylene signals to site 23: $\delta_H = 2.63$ & 2.72 ppm and $\delta_C = 38.8$ ppm.

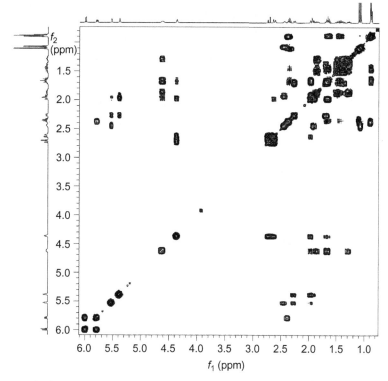

FIG. 3.2.5 The 2-D ^{1}H-^{1}H COSY NMR spectrum of lovastatin in chloroform-d.

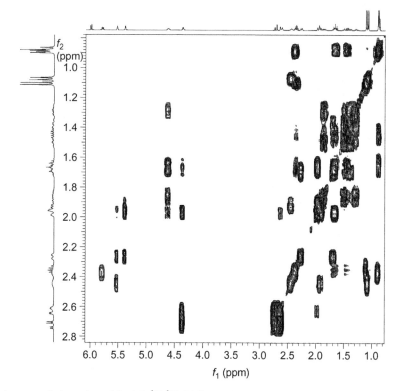

FIG. 3.2.6 An expanded portion of the 2-D ^{1}H-^{1}H COSY NMR spectrum of lovastatin in chloroform-d.

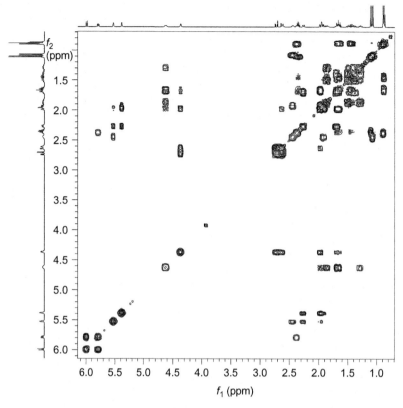

FIG. 3.2.7   Another expanded portion of the 2-D ^{1}H-^{1}H COSY NMR spectrum of lovastatin in chloroform-d.

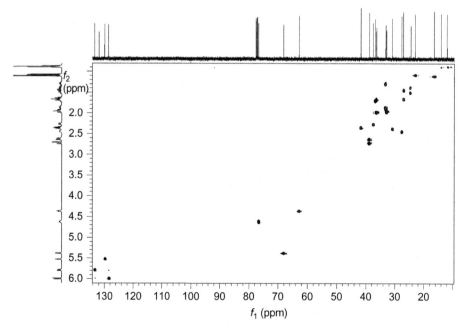

FIG. 3.2.8   The 2-D ^{1}H-^{13}C HSQC NMR spectrum of lovastatin in chloroform-d.

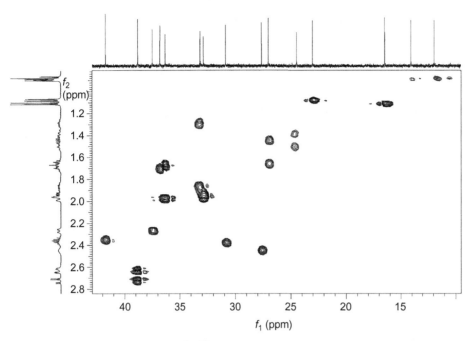

FIG. 3.2.9 An expanded portion of the 2-D ^{1}H-^{13}C HSQC NMR spectrum of lovastatin in chloroform-*d*.

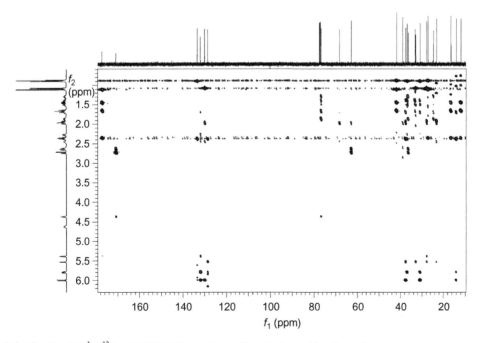

FIG. 3.2.10 The 2-D ^{1}H-^{13}C gHMBC NMR spectrum of lovastatin in chloroform-*d*.

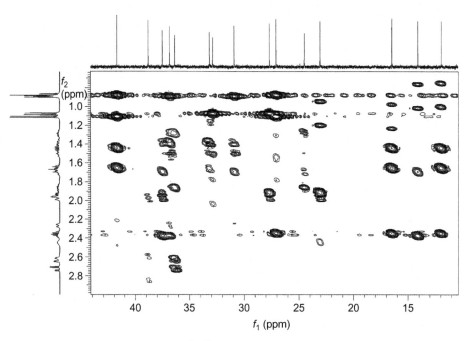

FIG. 3.2.11   An expanded portion of the 2-D ^{1}H-^{13}C gHMBC NMR spectrum of lovastatin in chloroform-d.

A methine ^{1}H signal in the midfield range of the HSQC (near the benchmark shift for aliphatic group next to oxygen of $\delta_H = 4$ ppm and $\delta_C = 60$ ppm) at 4.36 ppm shares COSY correlations with the H23 signals at 2.63 & 2.72 ppm. We assign this ^{1}H signal to the methine group bearing the hydroxyl at site 22: $\delta_H = 4.36$ ppm and $\delta_C = 62.8$ ppm. The site 22 hydroxyl group is likely responsible for the ^{1}H signal observed at $\delta_H = 2.61$ ppm, overlapping with the H23 signal at 2.63 ppm and making the integral of this region closer to two than one.

The H22 signal at 4.36 ppm also shares a pair of COSY cross peaks with methylene ^{1}H signals at 1.68 & 1.97 ppm. We assign the methylene ^{1}H signals to those of site 21 and so we write for site 21: $\delta_H = 1.68$ & 1.97 ppm and $\delta_C = 36.3$ ppm. The H21 signal at 1.97 ppm correlates weakly in the COSY spectrum with the hydroxyl signal at 2.61 ppm, confirming our identification of the site 22 hydroxyl ^{1}H signal.

The H21 signals at 1.68 & 1.97 ppm correlate with other ^{1}H signals in the COSY spectrum. The H21 signals share COSY cross peaks with a midfield methine ^{1}H signal at 4.62 ppm and not just with the H22 signal at 4.36 ppm. We assign this methine ^{1}H signal to site 20: $\delta_H = 4.62$ ppm and $\delta_C = 76.6$ ppm.

We are able to continue making assignments as we move from site 20 down into the remainder of the molecule. The COSY spectrum contains a pair of methylene ^{1}H signals at 1.28 & 1.86 ppm that correlate with the H20 signal at 4.62 ppm. This methylene group must be assigned to site 19: $\delta_H = 1.28$ & 1.86 ppm and $\delta_C = 33.1$ ppm.

The H19 signals at 1.28 & 1.86 ppm gratifyingly share COSY cross peaks with a fresh set of methylene ^{1}H signals at 1.39 & 1.50 ppm. The methylene ^{1}H signals at 1.39 & 1.50 ppm

TABLE 3.2.1   Predicted Multiplicities for the ^{1}H's of Lovastatin

Site in Molecule	Expected Multiplicity
1	d^3
2	td, dt
3	dqd^2
4	d
5	d
6	d^2
7	dqd
8	d^4
9	d^2
12	dqd
13	$2 \times dq$
14	d^2
15	d
16	d
17	d
18	$2 \times d^4$
19	$2 \times d^4$
20	d^4
21	$2 \times d^3$
22	d^4
22 (OH)	s
23	$2 \times d^2$

TABLE 3.2.2   Numbered Sites of Lovastatin Grouped by Group Type

Group Type	Site Number
CH_3 (methyl)	14, 15, 16, 17
CH_2 (methylene)	2, 13, 18, 19, 21, 23
CH (methine)	1, 3, 4, 5, 6, 7, 8, 9, 12, 20, 22
C_{np} (nonprotonated)	10, 11, 24

TABLE 3.2.3 ^{1}H and ^{13}C NMR Signals of Lovastatin Listed by Group Type

^{1}H Signal (ppm)	^{13}C Signal (ppm)	Group Type
–	177.1	Nonprotonated
–	170.7	Nonprotonated
5.79	133.3	Methine
–	131.8	Nonprotonated
5.52	129.8	Methine
5.99	128.5	Methine
4.62	76.6	Methine
5.38	68.1	Methine
4.36	62.8	Methine
2.35	41.7	Methine
2.63 & 2.72	38.8	Methylene
2.27	37.5	Methine
1.70	36.8	Methine
1.68 & 1.97	36.3	Methylene
1.28 & 1.86	33.1	Methylene
1.93 & 1.95[a]	32.8	Methylene
2.38	30.9	Methine
2.45	27.6	Methine
1.44 & 1.66	27.0	Methylene
1.39 & 1.50	24.5	Methylene
1.07	23.0	Methyl
1.11	16.4	Methyl
0.89[b]	14.1	Methyl
0.88[b]	11.9	Methyl

[a]The two ^{1}H's are not equivalent and it appears they show distinct but similar chemical shifts such that their resonances show pronounced nonfirst-order splitting effects.
[b]The HSQC spectrum fails to clearly show the two most upfield methyl correlations—we use the gHMBC spectrum and average the positions of the two HMQC cross peaks that are $^1J_{CH}$-split in the f_2 (^{1}H) dimension.

are attributed to site 18 and we write for site 18: $\delta_H = 1.39$ & 1.50 ppm and $\delta_C = 24.5$ ppm. Although there is significant overlap of the ^{1}H resonances for these methylene groups making it challenging to verify our newest assignments, we can obtain a small amount of confirmation by noting the presence of a gHMBC cross peak between one of the H19 signals (1.86 ppm) and the newly assigned C18 signal (24.5 ppm).

Having assigned the site 18 methylene NMR signals we have just the site 2 methylene group remaining. By process of elimination we assign site 2 as follows: $\delta_H = 1.93$ & 1.95 and $\delta_C = 32.8$ ppm.

We now turn to the lower portion of lovastatin as it is shown in Fig. 3.2.1. We have already identified the C10 signal at 131.8 ppm. We can easily identify the three downfield $^1H/^{13}C$ signals from sites 4, 5, and 6, which are the protonated sites of the conjugated carbon-carbon double bonds. Examination of the downfield portion of the COSY spectrum of lovastatin reveals a pair of cross peaks between ^1H signals at 5.99 and 5.79 ppm. The second-most downfield ^1H signal is a doublet of doublets at 5.79 ppm. We assign this signal to H6 and the plain doublet at 5.99 ppm we assign to H5. We write for site 6: $\delta_H = 5.79$ and $\delta_C = 133.3$ ppm and for site 5: $\delta_H = 5.99$ and $\delta_C = 128.5$ ppm. The H4 signal is the only remaining downfield HSQC cross peak and so we assign site 4 as follows: $\delta_H = 5.52$ ppm and $d_C = 129.8$ ppm.

In the COSY spectrum, the methine ^1H signal at 2.38 ppm correlates with the H6 signal at 5.79 ppm. Because we have already assigned the methine ^1H signal at 2.35 ppm to the remote H12, we are confident that the H7 signal is that observed at 2.38 ppm, as there are no other signals which are near 2.35–2.38 ppm on the ^1H chemical shift axis. We write the following for site 7: $\delta_H = 2.38$ ppm and $\delta_C = 30.9$ ppm.

The methyl ^1H signal at 0.89 ppm correlates strongly in the gHMBC spectrum with the C7 signal at 30.9 ppm. This must be the methyl ^1H signal from site 17 and so have assigned the site 17 ^1H signal and also confirmed our assignment of C7. We write for site 17: $\delta_H = 0.89$ ppm and $\delta_C = 14.1$ ppm. The H17 signal at 0.89 ppm shares a confirming gHMBC cross peak with the C6 signal at 133.3 ppm.

By the process of elimination we can assign the last methyl NMR signals as being those from site 16: $\delta_H = 1.07$ ppm and $\delta_C = 23.0$ ppm.

Both the H16 signal at 1.07 ppm and the H4 signal at 5.52 ppm should share gHMBC cross peaks with the C3 signal. We observe a methine ^{13}C signal at 27.6 ppm with these attributes. We write for site 3: $\delta_H = 2.45$ ppm and $\delta_C = 27.6$ ppm. The putative H3 resonance at 2.45 ppm shares gHMBC cross peaks with the signals of C4 at 129.8 ppm and C16 at 23.0 ppm, confirming our site 3 assignment. The H3 signal at 2.45 ppm also shares a gHMBC correlation with the already assigned C10 signal at 131.8 ppm, also consistent with our site 3 assignment.

There are three methine groups at sites 1, 8, and 9 that remain to be assigned. The site 1 methine group is alpha to oxygen and so we expect its HSQC cross peak to be found in the midfield region near the ^{13}C chemical shift of 60 ppm. The sole remaining methine signal in the midfield region of the HSQC spectrum is assigned to site 1 as follows: $\delta_H = 5.38$ ppm and $\delta_C = 68.1$ ppm. The H1 signal at 5.38 ppm shares gHMBC cross peaks with the signals of C10 at 131.8 ppm and C3 at 27.6 ppm. We note that the H1 signal at 5.38 ppm correlates with the H2 signals at 1.93 & 1.95 ppm in the COSY spectrum. Interestingly, the H1 signal is not observed to correlate with the C2 signal at 32.8 ppm in the gHMBC spectrum. This may be the result of setting the plot threshold at a high value to minimize t_1 ridges.

The H1 signal at 5.38 ppm also shares a COSY cross peak with a methine ^1H signal at 2.27 ppm. This signal must correspond to H9 and so we write for site 9: $\delta_H = 2.27$ ppm and $\delta_C = 37.5$ ppm.

The H9 is expected to couple to the site 8 methine ^1H. H9 generates the last unassigned methine ^1H signal at 1.70 ppm. We observe a COSY cross peak between the signals of H9 at 2.27 ppm and H8 at 1.70 ppm, allowing us to write for site 8: $\delta_H = 1.70$ ppm and $\delta_C = 36.8$ ppm. We confirm our site 8 assignment by noting that the H8 signal at 1.70 ppm shares a strong gHMBC cross peak with the C17 signal at 14.1 ppm. This cross peak is predicted to be large because of the *trans*-$^3J_{CH}$ expected to occur between the adjacent axial groups on the lower-right-most ring of lovastatin.

3.3 GIBBERELLIC ACID IN DIMETHYL SULFOXIDE-D_6

Gibberellic acid is a 19-carbon molecule containing five rings. Two pairs of these rings share three common carbons and three of the five rings are five-membered. The structure of gibberellic acid appears in Fig. 3.3.1. The molecule also has two carbonyls (one carboxylic acid, one lactone), two widely separated carbon-carbon double bonds, and two hydroxyl groups. The index of hydrogen deficiency is nine because of the five rings and four double bonds, so the empirical formula is, accounting for the carboxyl, ester, and two hydroxyls, $C_{19}H_{22}O_6$. The 1-D ^1H NMR spectrum is shown in Fig. 3.3.2 and the 1-D ^{13}C NMR spectrum is shown in Fig. 3.3.3. The 2-D ^1H-^1H COSY NMR spectrum appears in Fig. 3.3.4, the 2-D ^1H-^{13}C HSQC NMR spectrum is found in Fig. 3.3.5, and the 2-D ^1H-^{13}C gHMBC NMR spectrum is shown in Fig. 3.3.6.

Construction of the molecular model for this particular compound assists us in understanding the NMR spectra of gibberellic acid because it allows us to understand couplings that depend on molecular conformation—not just how molecular orbitals overlap for a given local conformation, but also how molecular flexibility will hinder our ability to observe couplings involving dynamic dihedral angles. The ring containing carbons 7, 8, 9, 10, 11, and 12 appears able to adopt more than one low-energy conformation. The ring below and to the right of the aforementioned ring (as shown in Fig. 3.3.1) may also exhibit a limited amount of conformational variability. Variation of conformation suppresses many of the COSY and gHMBC cross peaks we would normally expect to observe, thus making our task of assigning NMR signals to molecular sites more challenging. When a molecule samples a number of nearly degenerate conformational states on the NMR timescale, the NMR experiments we carry out that depend on low frequency interactions—i.e., J-couplings—will often fail to generate useful results. Variation of couplings over the course of a single pass through the NMR pulse sequence will serve to average out phase coherence, giving signal cancelation.

We are cautious in predicting the multiplicities of the ^1H signals of gibberellic acid, for example, writing d^4 instead of qd or tt. Examination of the structure shows us that none of

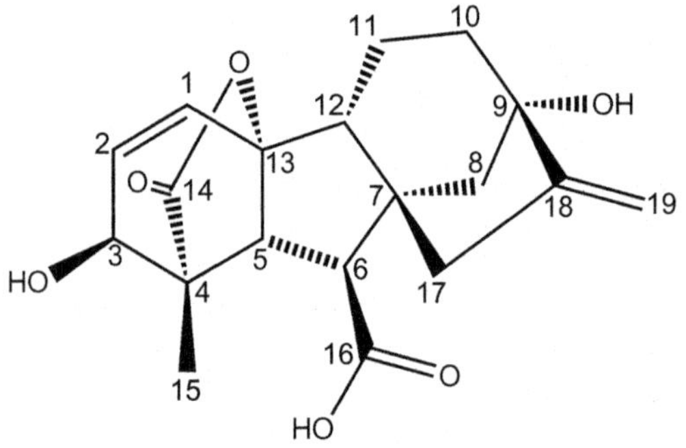

FIG. 3.3.1   The structure of gibberellic acid.

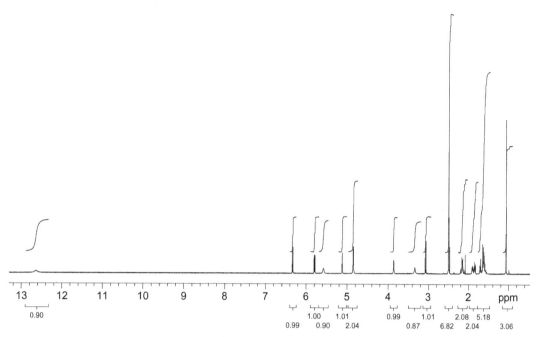

FIG. 3.3.2   The 1-D ^{1}H NMR spectrum of gibberellic acid in dimethyl sulfoxide-d_6.

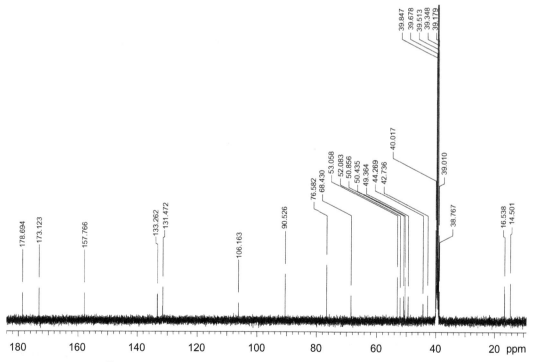

FIG. 3.3.3   The 1-D ^{13}C NMR spectrum of gibberellic acid in dimethyl sulfoxide-d_6.

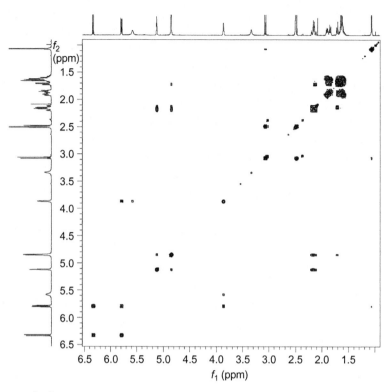

FIG. 3.3.4   The 2-D ^{1}H-^{1}H COSY NMR spectrum of gibberellic acid in dimethyl sulfoxide-d_6.

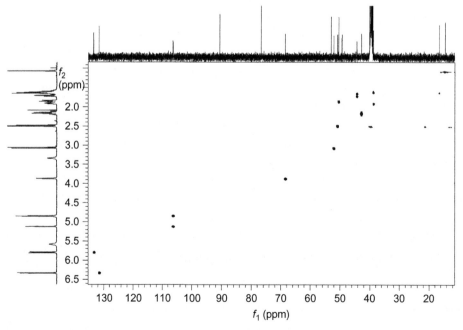

FIG. 3.3.5   The 2-D ^{1}H-^{13}C HSQC NMR spectrum of gibberellic acid in dimethyl sulfoxide-d_6.

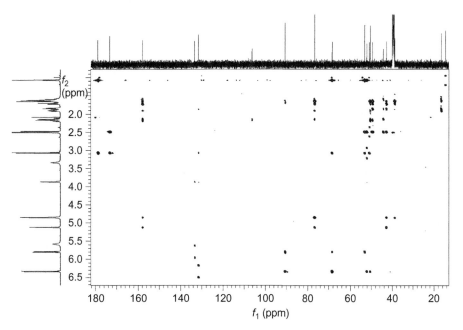

FIG. 3.3.6 The 2-D ^{1}H-^{13}C gHMBC NMR spectrum of gibberellic acid in dimethyl sulfoxide-d_6.

the six-membered rings of the molecule adopt the chair conformation that is the norm in less constrained systems. The three five-membered rings are more rigid than other five-membered rings in molecules we may have assigned previously, again because of the highly constrained nature of the molecule, with its five rings and two carbon-carbon double bonds packed into only a 19-carbon package. This highly constrained system may also fail to generate some of the gHMBC cross peaks we expect to arise from $^2J_{CH}$'s because of bond angle distortion due to the constraints imposed by hydrogen deficiency.

We predict the multiplicities of the ^{1}H resonances of gibberellic acid and place these predictions in Table 3.3.1. We show our accounting of the various carbon types of gibberellic acid in Table 3.3.2. While our solvent is aprotic, it is also hygroscopic, so and not as free of labile protons as we might hope. We see not only the DMSO solvent signal the 1-D ^{1}H spectrum at 2.5 ppm, but also a broad peak at 3.34 ppm from water that has made its way into our DMSO. One of our solute ^{1}H signals is near that from the residual protonated solvent (present as DMSO-d_5,h_1), but the HSQC cross peak is easy to differentiate from the solvent peak. In the ^{13}C spectrum, the DMSO signal is dominated by the perdeuterated species, and so the ^{13}C line of the solvent is split by three deuterons to give a seven-line multiplet. A careful reading of the HSQC spectrum allowed us to populate Table 3.3.3, but already we can see consequences of conformational variation or some other source of chaos, for the correlation between what we must conclude are methylene ^{1}H signals in the 1.60–1.67 ppm range and the ^{13}C signal at 16.5 ppm is weak. The gHMBC spectrum is little if any help in resolving this question. Because only one set of signals is difficult to observe for the five methylene groups, we can determine its shifts by process of elimination. That is, we use the integrals to recognize that five ^{1}H signals must fall into the range from 1.55 to 1.75 ppm. We will find the hydroxyl signals

TABLE 3.3.1   Predicted Multiplicities for the ^{1}H's of Gibberellic Acid

Site in Molecule	Expected Multiplicity
1	d
2	d
3	d
3 (OH)	s
5	d
6	d
8	2×d
9 (OH)	s
10	2×d^3
11	2×d^4
12	d^2
15	s
16 (CO$_2$H)	s
17	2×d
19	2×d

TABLE 3.3.2   Carbons of Gibberellic Acid

Type of Carbon	Site Number
CH$_3$ (methyl)	15
CH$_2$ (methylene)	8, 10, 11, 17, 19
CH (methine)	1, 2, 3, 5, 6, 12
C$_{np}$ (nonprotonated)	4, 7, 9, 13, 14, 16, 18

elsewhere, but it is good to keep the hydroxyl ^{1}H signals in mind as we attempt to locate the missing methylene ^{1}H signals.

The ^{1}H signal at 12.68 ppm is uniquely downfield and is assigned to the carboxyl acid ^{1}H signal from site 16: $\delta_H = 12.86$ ppm.

In the downfield ^{13}C region of the HSQC spectrum, we observe a ^{13}C signal at 106.1 ppm that correlates with two ^{1}H signals at 4.85 & 5.12 ppm. We assign these NMR signals to the methylene group at site 19: $\delta_H = 4.85$ & 5.12 ppm and $\delta_C = 106.1$ ppm.

Although ^{1}H's on sp^2-hybridized carbon atoms often do not have signals that generate gHMBC cross peaks with ^{13}C's two bonds away (because of the 120 degrees bond angle which serves to minimize orbital overlap and hence minimize coupling), we observe significant gHMBC cross peak intensity between the H19 signals at 4.85 & 5.12 ppm and the nonprotonated ^{13}C resonance at 157.8 ppm. This downfield ^{13}C chemical shift can only be attributed to site 18: $\delta_C = 157.8$ ppm.

TABLE 3.3.3 ^{1}H and ^{13}C NMR Signals of Gibberellic Acid Listed by Group Type

^{1}H Signal (ppm)	^{13}C Signal (ppm)	Group Type
–	178.7	Nonprotonated
–	173.1	Nonprotonated
–	157.8	Nonprotonated
5.58	133.3	Methine
6.33	131.5	Methine
4.85 & 5.12	106.2	Methylene
–	90.5	Nonprotonated
–	76.6	Nonprotonated
3.87	68.4	Methine
–	53.1	Nonprotonated
3.07	52.1	Methine
2.48	50.9	Methine
1.87	50.4	Methine
–	49.4	Nonprotonated
1.70 & 1.74	44.3	Methylene
2.14 & 2.18	42.7	Methylene
1.67 & 1.93	38.8	Methylene
1.61 & 1.64[a]	16.5	Methylene
1.07	14.5	Methyl
12.68	–	Carboxyl/phenol
5.58	–	Hydroxyl
4.85	–	Hydroxyl

[a]*These chemical shifts obtained from the 1-D ^{1}H NMR spectrum.*

The well-resolved H19 signals also correlate with a more upfield ^{13}C resonance from another nonprotonated carbon site. This more upfield ^{13}C resonance at 76.6 ppm is assigned to site 9: $\delta_C = 76.6$ ppm.

We would like to use an argument that the H19 *trans* to C9 will couple more strongly with C9 than will the H19 *cis* to C9. Unfortunately, the integral of the ^{1}H resonance at 4.85 ppm is 2.04 in Fig. 3.3.2, indicating that a hydroxyl resonance is also found at 4.85 ppm, and thus complicating any attempt to compare gHMBC cross peak intensity between the signals of the H19's and C9. In the gHMBC spectrum, the cross peaks between the H19 signals at 4.85 & 5.12 ppm and that of C9 at 76.6 ppm appear to be of equal intensity.

We can identify the H1 and H2 signals easily in the COSY spectrum by looking for the most downfield pair of cross peaks. We expect that H2 is coupled to not just H1 but also H3. The ^{1}H multiplet at 5.79 ppm is observed to be a doublet of doublets, while the ^{1}H resonance at 6.33 is a doublet. We assign the d^2 signal to site 2: $\delta_H = 5.79$ ppm and $\delta_C = 133.3$ ppm. The other ^{1}H signal we assign to site 1: $\delta_H = 6.33$ ppm and $\delta_C = 131.5$ ppm.

The H2 signal shares a strong COSY cross peak with the methine ^{1}H resonance at 3.87 ppm. We assign this new methine ^{1}H signal to site 3: $\delta_H = 3.87$ ppm and $\delta_C = 68.4$ ppm. The H3 signal also correlates with a hydroxyl ^{1}H signal at 5.58 ppm in the COSY spectrum, we assign the site 3 hydroxyl as follows: $\delta_H = 5.58$ ppm. We assign to the site 9 hydroxyl ^{1}H signal at 4.85 ppm by elimination.

We now assign the sole methyl group in our molecule. The ^{1}H signal at 1.07 ppm is assigned to the H15 and so we write for site 15: $\delta_H = 1.07$ ppm and $\delta_C = 14.5$ ppm.

The H15 signal at 1.07 ppm shares a gHMBC cross peak with the nonprotonated ^{13}C signal at 178.7 ppm, with the nonprotonated ^{13}C signal at 53.1 ppm, and with the methine ^{13}C signal at 52.1 ppm. We assign the most downfield ^{13}C signal to the lactone carbonyl at site 14: $\delta_C = 178.7$ ppm. We assign the other nonprotonated ^{13}C signal to site 4: $\delta_C = 53.1$ ppm. The methine ^{13}C signal that correlates with the H15's we assign to site 5: $\delta_H = 3.07$ ppm and $\delta_C = 52.1$ ppm.

By the process of elimination, we identify the site 16 ester carbonyl signal as being the other ^{13}C resonance with a chemical shift greater than 170 ppm. We write for site 16: $\delta_C = 173.1$ ppm.

The H5 signal at 3.07 ppm shares a strong COSY cross peak with a methine ^{1}H signal at 2.48 ppm. We assign this signal to H6 since H5 and H6 together form an isolated spin system. We write for site 6: $\delta_H = 2.48$ ppm and $\delta_C = 50.9$ ppm. The H6 signal at 2.48 ppm shares a strong gHMBC cross peak with the C16 signal at 173.1 ppm, thus neatly confirming both the site 6 and site 16 assignments.

Because we have assigned all the other methine carbons, we can by elimination assign the last set of methine NMR signals to site 12: $\delta_H = 1.87$ ppm and $\delta_C = 50.4$ ppm.

Of the two nonprotonated sp^3-hybridized carbon sites still to be assigned, C13 is alpha to an oxygen while C7 is bound only to carbon. The H2 signal at 5.79 ppm is found to correlate with the ^{13}C signal at 90.5 ppm, which is the more downfield of the two unassigned nonprotonated ^{13}C signals. We assign this ^{13}C signal to site 13: $\delta_C = 90.5$ ppm. The H1 signal at 6.33 ppm shares a confirming gHMBC cross peak with the C13 signal at 90.5 ppm.

The chemical shift of the C7 signal must be 49.4 ppm, because that is the shift of the last unassigned, nonprotonated ^{13}C resonance. We write for site 7: $\delta_C = 49.4$ ppm. We confirm this assignment by noting that the H6 signal at 2.48 ppm shares a strong gHMBC cross peak with the C7 signal at 49.4 ppm.

We now turn to the NMR signals of four methylene groups. The four methylenes are in the top-right portion of the molecule as shown in Fig. 3.3.1 and are found at sites 8, 10, 11, and 17. We observe gHMBC cross peaks between the H19 signals at 4.85 & 5.12 ppm and the nonprotonated C18 signal at 157.8 ppm, the nonprotonated C9 signal at 76.6 ppm, and a methylene ^{13}C signal at 42.7 ppm. We assign this methylene ^{13}C signal to site 17: $\delta_H = 2.14$ & 2.18 ppm and $\delta_C = 42.7$ ppm. We obtain a lukewarm confirmation of this assignment by noting that the H17 signals at 2.14 & 2.18 ppm share gHMBC cross peaks with the C6 signal at 50.9 ppm. This confirmation would be more significant if the H8's were not equally positioned to generate the same gHMBC correlation with C6.

Site 11 can be differentiated from sites 8 and 10 by noting that sites 8 and 10 are beta to the oxygen atom bound to C9. We expect that C11 will therefore be found upfield from C8 and C10, allowing us to assign site 11 as follows: $\delta_H = 1.61$ & 1.64 ppm and $\delta_C = 16.5$ ppm.

The H12 signal at 1.87 ppm shares a COSY cross peak with methylene ^{1}H signals at 1.70 & 1.74 ppm, and also with one of the two H11 methylene signals at 1.61 & 1.64 ppm. The methylene ^{1}H signals at 1.70 & 1.74 ppm could be from site 10 or from site 8, as both are equally removed from H12.

The H6 resonance at 2.48 ppm shares gHMBC cross peaks with the signals of C16 at 173.1 ppm, C4 at 53.1 ppm, C5 at 52.1 ppm, C7 at 49.4 ppm, C17 at 42.7 ppm, and a methylene ^{13}C signal at 44.3 ppm. Because H6 is unlikely to couple to C10, five bonds away, we assign this methylene ^{13}C signal instead to site 8: $\delta_H = 1.70$ & 1.74 ppm and $\delta_C = 44.3$ ppm.

By elimination, our last methylene group is site 10: $\delta_H = 1.67$ & 1.93 ppm and $\delta_C = 38.8$ ppm. The well-resolved H10 signal at 1.93 ppm is observed to share gHMBC cross peaks with the signals of C18 at 157.8 ppm, C9 at 76.6 ppm, C12 at 50.4 ppm, and C11 at 16.5 ppm.

The flexibility of the site 10 and site 11 methylenes reduces the number of cross peaks we observe involving the signals from the spins at these sites. If we tally the number of cross peaks in the gHMBC involving the ^{1}H signals of the methylenes, we see that the H8 signals participate in seven cross peaks, while the H10 and H11 signals each participate in only four cross peaks.

The H12 signal at 1.87 ppm shares gHMBC cross peaks with the signals of C7 at 49.4 ppm, C8 at 44.3 ppm, C17 at 42.7 ppm, and C11 at 16.5 ppm, but not C10 because of the flexibility of sites 10 and 11. On the other hand, the hydroxyl ^{1}H signal of site 9 likely shares a correlation with C9 and C10, and possibly with C18 and C17 as well, although the overlap of the site 9 hydroxyl ^{1}H signal and that of one of the H19's makes these cross peaks less convincing in their confirmation of the overall correctness of our assignment of sites 8, 10, and 11.

A point to ponder is: Why is the C13 signal involved in so few gHMBC cross peaks? It is noted that neither the signals of H5 nor H12 shares a gHMBC cross peak with C13 signal. Perhaps bond angles involving H12-C12-C13 and H5-C5-C13 are both significantly greater than the nominal 109.5 degrees bond angle we expected based on sp^3-hybridization.

3.4 NARINGIN HYDRATE IN METHANOL-D$_4$

Naringin is a glycoside (one or more sugars plus something else) derived from naringenin, which is a significant component of grapefruit and other citrus fruits. The structure of naringin is shown in Fig. 3.4.1. Naringin is sometimes confused with naringenin, which lacks the lower-left sugar rings and instead has a hydroxyl group at site 7. Naringin has five rings, two of which are aromatic and one that is heterocyclic. The other two rings are six-membered cyclic monosaccharides. Naringin has two phenolic hydroxyl groups, and its sugar rings bear six hydroxyl groups. All eight hydroxyl signals exchange with the labile deuterons in the solvent and, because of the large molar excess of solvent to solute, spend the majority of their time on the hydroxyl group of methanol, and so appear in the 1-D ^{1}H NMR spectrum (Fig. 3.4.2) at 4.90 ppm, the chemical shift of the residual protonated solvent. The integral of this peak should be at minimum 10 (eight hydroxyl groups plus two ^{1}H's from the hydrate)— it comes in gratifyingly at 12-something. The empirical formula of naringin is $C_{27}H_{32}O_{14}$.

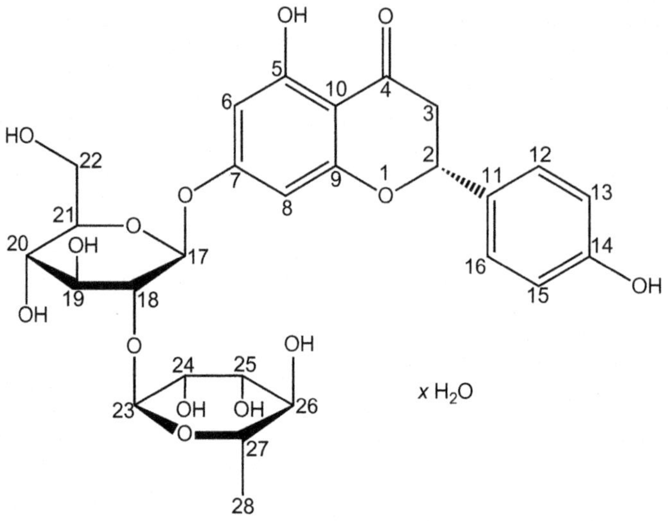

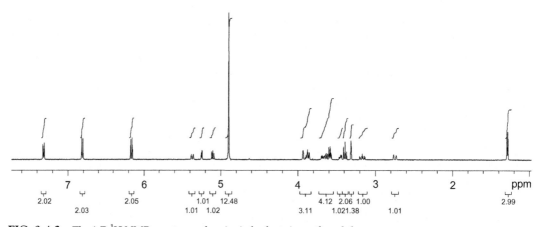

FIG. 3.4.1   The structure of naringin hydrate.

FIG. 3.4.2   The 1-D ^{1}H NMR spectrum of naringin hydrate in methanol-d_4.

A small amount of naringin was dissolved in methanol-d_4 and used to generate the spectra appearing in this section. The 1-D ^{1}H NMR spectrum obtained from the sample appears in Fig. 3.4.2. The 1-D ^{13}C NMR spectrum of naringin is shown in Fig. 3.4.3. Some of the ^{13}C signals are doubled, and so we have the significant challenge of ensuring that we correctly differentiate between two ^{13}C signals and one ^{13}C signal that is split. Shimming is not the cause of the ^{13}C peak doubling, because if it were, all the chemical shift differences recorded by the peak picking algorithm would be similar. The difference in the observed ^{13}C chemical shifts, $\Delta\delta$, is as great as 0.201 ppm, with three other ^{13}C signals displaying $\Delta\delta$'s near 0.070 ppm. With this molecule, the doubling of the signals is variable from site to site. It is believed that the molecule may slowly, on the NMR timescale, sample two distinct conformations, perhaps through rotation of the C7-O-C17 ether linkage. A second possibility for the doubling of the

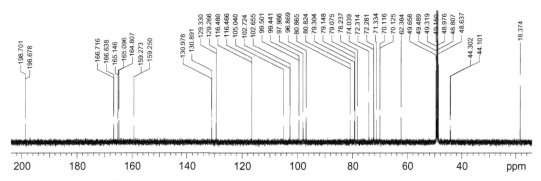

FIG. 3.4.3 The 1-D ^{13}C NMR spectrum of naringin hydrate in methanol-d_4.

^{13}C signals is that the two sugar rings are sometimes attached to the naringenin precursor at site 5 instead of at site 7. A third possibility is that the stereochemistry at site 2 may vary (searching online in August 2016, the author finds structures both opposite of the original chemical catalog structure and also with no specified stereochemistry at site 2—this ambiguity is in sharp contrast with the complete agreement found for the two sugar rings in the molecule). Whatever the cause, we diligently note the presence of this complication, taking care as we ensure we identify the correct number of ^{13}C signals of each type.

Rotational symmetry about the C2-C11 bond reduces the number of distinct NMR signals we observe. Naringin's NMR spectra feature 15 methine signal pairs accounting for 17 ^{1}H/^{13}C carbon pairs. As far as NMR assignment problems of sugars go, naringin is rather tame, but the midfield region of the ^{1}H chemical shift range still has enough resonance overlap to familiarize us with the treacherous territory in the spin systems of cyclic sugars that lie between the anomeric and terminal methylene/methyl groups. The 2-D ^{1}H-^{1}H COSY NMR spectrum is found in Fig. 3.4.4. A 2-D ^{1}H-^{1}H TOCSY NMR spectrum obtained with a mixing time of 120 ms is shown in Fig. 3.4.5. The TOCSY spectrum is especially useful for the examination of the ^{1}H spin systems of cyclic sugars, as the COSY spectrum's ability to propagate the progress of the assignment is hindered by midfield ^{1}H resonance overlap. We supplement homonuclear ^{1}H NMR information with heteronuclear spectra involving ^{13}C. The 2-D ^{1}H-^{13}C HSQC NMR spectrum of naringin appears in Fig. 3.4.6, and the 2-D ^{1}H-^{13}C gHMBC NMR spectrum is displayed in Fig. 3.4.7. In the midfield region, an expansion of the gHMBC spectrum appears in Fig. 3.4.8.

We examine the structure of naringin and predict the multiplicities of the ^{1}H signals at the various molecular sites, putting these predictions into Table 3.4.1. Examining the structure of naringin, we also determine the group type for the numbered molecular sites of naringin, putting our tabulations into Table 3.4.2. All of the ^{1}H and ^{13}C NMR signals attributable to the solute appear in Table 3.4.3, sorted by molecular site using the HSQC spectrum.

We are able to locate the NMR signal from the sole methyl group at site 28 immediately by finding the most upfield ^{1}H and ^{13}C resonances, which also show a correlation in the HSQC spectrum. We write for site 28: $\delta_H = 1.28$ ppm and $\delta_C = 18.4$ ppm. Assigning that much of the sugar ring NMR signals was easy, the rest of the assignment will be less so; we leave it for later.

There is only one carbonyl carbon in our molecule as well, so we can assign that to the most downfield ^{13}C resonance, writing for site 4: $\delta_C = 198.7$ ppm.

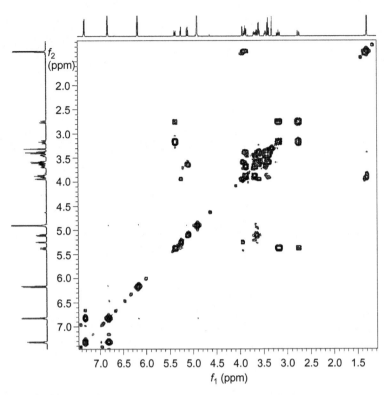

FIG. 3.4.4 The 2-D ^{1}H-^{1}H COSY NMR spectrum of naringin hydrate in methanol-d_4.

Moving to lower ppm values from the carbonyl ^{13}C resonance, we encounter four nonprotonated ^{13}C resonances at 166.7, 165.1, 164.8, and 159.3 ppm. Aromatic ^{13}C's bonded to oxygen are typically found to have a chemical shifts of at least 150 ppm. Naringin has these carbon chemical environments at sites 5, 7, 9, and 14. We might speculate that the C14 signal will be most upfield because site 14 does not experience the electron withdrawal expected at sites 5, 7, and 9, which are respectively *ortho*, *para*, and *ortho* to the electron-withdrawing C4 carbonyl.

The 1-D ^{1}H NMR spectrum contains three sets of downfield ^{1}H signals at 7.32, 6.82, and 6.16 ppm that each have an integral corresponding to two ^{1}H's in naringin. We were expecting only two sets from the symmetry-doubled sites of 12/16 and 13/15. Examination of the HSQC spectrum allows us to determine that the most upfield set of ^{1}H signals at ~6.16 ppm arise from two distinct methine groups, as the ^{1}H signals at 6.14 and 6.18 ppm clearly correlate with two ^{13}C resonances in the HSQC NMR spectrum. The COSY spectrum reveals that the ^{1}H signals at 7.32 ppm and those at 6.82 ppm are strongly correlated, assuring that we have identified the H12/16 and H13/15 signals. Although the 1-D ^{13}C NMR spectrum has hash marks from the peak picking algorithm that prevent us from identifying the doubly intense signals from C12/16 and C13/15, we can locate these double intensity peaks at the top (f_1 horizontal) of the HSQC and gHMBC spectra. The lone pairs on the site 14 hydroxyl group donate electron density to sites 13 and 15, which are *ortho* with respect to site 14, but not to sites 12 and 16, which are *meta* to site 14. Electron density donated from site 14's oxygen and

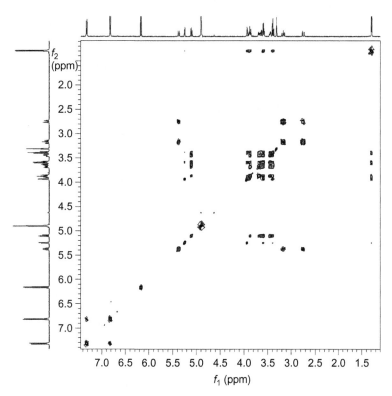

FIG. 3.4.5 The 2-D ^{1}H-^{1}H TOCSY NMR spectrum of naringin hydrate in methanol-d_4 obtained with a mixing time of 120 ms.

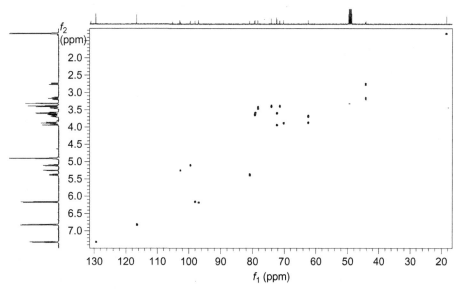

FIG. 3.4.6 The 2-D ^{1}H-^{13}C HSQC NMR spectrum of naringin hydrate in methanol-d_4.

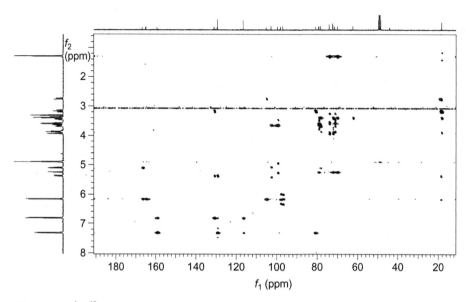

FIG. 3.4.7   The 2-D ^{1}H-^{13}C gHMBC NMR spectrum of naringin hydrate in methanol-d_4.

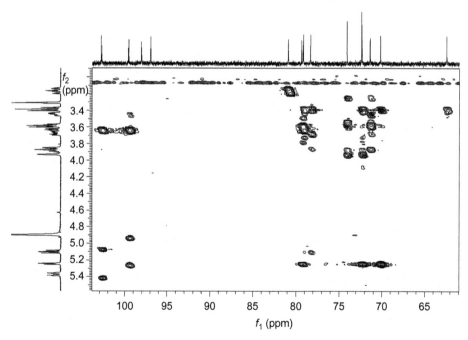

FIG. 3.4.8   An expanded portion of the 2-D ^{1}H-^{13}C gHMBC NMR spectrum of naringin hydrate in methanol-d_4.

TABLE 3.4.1 Predicted Multiplicities for the ^{1}H Resonances of Naringin

Site in Molecule	Expected Multiplicity
2	d^2
3	$2 \times d^2$
5 (OH)	s
6	s
8	s
12/16	d
13/15	d
14 (OH)	s
17	d
18	d^2
19	d^2
19 (OH)	s
20	d^2
20 (OH)	s
21	d^3
22	$2 \times d^2$
22 (OH)	s
23	d
24	d^2
24 (OH)	s
25	d^2
25 (OH)	s
26	d^2
26 (OH)	s
27	dq
28	d

accepted at sites 13/15 causes the site 13/15 spins to experience additional shielding, thereby lowering their NMR resonant transition frequencies to smaller chemical shift values (upfield). We are able to differentiate the NMR signals of sites 12/16 from those of sites 13/15 because we are able to assign the upfield ^{1}H/^{13}C NMR shift pair to sites 13/15: $\delta_H = 6.82$ ppm and $\delta_C = 116.5$ ppm. We assign the more downfield set of NMR signals to sites 12/16: $\delta_H = 7.32$ ppm and $\delta_C = 129.3$ ppm.

TABLE 3.4.2 Numbered Molecular Sites of Naringin

Group Type	Site Number
CH_3 (methyl)	28
CH_2 (methylene)	3, 22
CH (methine)	2, 6, 8, 12/16, 13/15, 17, 18, 19, 20, 21, 23, 24, 25, 26, 27
C_{np} (nonprotonated)	4, 5, 7, 9, 10, 11, 14
Oxygen	1

TABLE 3.4.3 ^{1}H and ^{13}C NMR Signals of Naringin Listed by Group Type

[1]H Signal (ppm)	[13]C Signal (ppm)	Group Type
–	198.7	Nonprotonated
–	166.7	Nonprotonated
–	165.1	Nonprotonated
–	164.8	Nonprotonated
–	159.3	Nonprotonated
–	130.9	Nonprotonated
7.32	129.3	Methine
6.82	116.5	Methine
–	105.0	Nonprotonated
5.24	102.7	Methine
5.10	99.5	Methine
6.14	98.0	Methine
6.18	96.9	Methine
5.37	80.8	Methine
3.64	79.3	Methine
3.58	79.1	Methine
3.44	78.2	Methine
3.39	74.0	Methine
3.93	72.31	Methine
3.59	72.28	Methine
3.38	71.3	Methine
3.88	70.1	Methine
3.69 & 3.86	62.4	Methylene
2.76 & 3.16	44.2	Methylene
1.28	18.4	Methyl

We expect a large trans-$^3J_{CH}$ between the H12/16's and C14 and so we look at the gHMBC spectrum to see which of the four nonprotonated ^{13}C signals in the 159–167 ppm range correlates with the ^{1}H signal at 7.32 ppm. We observe a strong gHMBC cross peak between the H12/16 signal at 7.32 ppm and the nonprotonated ^{13}C signal at 159.3 ppm, allowing us to write for site 14: $\delta_C = 159.3$ ppm. While this assignment confirms our earlier prediction that the C14 signal should be more upfield from the signals of the carbonyl-conjugated C5, C7, and C9, we should not place too much faith in this type of argument. We rely on these resonance predictions when we lack more definitive means of assigning NMR signals to molecular sites.

Using the same trans-$^3J_{CH}$ interaction, we can use the H13/15 signals to assign site 11. In the gHMBC spectrum, the H13/15 signal at 6.82 ppm shares its strongest cross peak with the nonprotonated ^{13}C signal at 130.9 ppm. We assign this signal to site 11: $\delta_C = 130.9$ ppm. The H13/15 signals at 6.82 ppm also correlate with the signals of C14 at 159.3 ppm, C12/16 at 129.3 ppm, and C15/13 at 116.5 ppm. This third cross peak arises from a physical coupling crossover and should not be mistaken for an HMQC-type $^1J_{CH}$ interaction: The gHMBC cross peak shared between the ^{1}H shift of 6.82 ppm and the ^{13}C shift of 116.5 ppm arises from $^3J_{CH}$'s between H13 and C15, and also between H15 and C13.

The H12/16 signals at 7.32 ppm correlate with the C14 signal at 159.3 ppm, with the C16/12 signal at 129.3 ppm, with the C13/15 signal at 116.5 ppm, and with a methine ^{13}C signal at 80.8 ppm. We assign this methine ^{13}C signal to site 2: $\delta_H = 5.37$ ppm and $\delta_C = 80.8$ ppm. The chemical shift of the site 2 signals is consistent with the alpha oxygen atom.

The newly found H2 signal at 5.37 ppm shares a pair of COSY cross peaks with the methylene ^{1}H signals at 2.76 & 3.16 ppm. We assign these methylene signals to site 3: $\delta_H = 2.76$ & 3.16 ppm and $\delta_C = 44.2$ ppm. The site 3 ^{13}C signal is the most split of all the ^{13}C signals. The unequal intensity of the two lines associated with the C3 signal suggests two distinct populations and adds weight to our third argument that the variable ^{13}C signal doubling is from having an R- versus and S-stereocenter at site 2 of the molecule. This is an easy to understand example of epimers, which are molecules with more than two stereocenters that are identical example at one stereocenter only. Naringin has eleven stereocenters, with five apiece on the two sugar rings. We can imagine that inversion at site 2 might affect how the rings interact with the top-right portion of the molecule as it is shown in Fig. 3.4.1. Fortunately, the two apparent epimers have sufficiently similar ^{1}H and ^{13}C NMR behavior.

We are able to confirm the assignment of site 3 by noting a pair of cross peaks in the gHMBC spectrum between the H3 signals at 2.76 & 3.16 ppm and the C4 signal, but in the gHMBC spectrum the f_1 spectral window was set too small by about 10 ppm on the downfield (higher ppm value) side. This error probably occurred because the author saw the site 1 oxygen and the site 4 carbonyl near each other and mistakenly believed that the carbonyl was a lactone and so would generate a signal in the 165–175 ppm range. The author resolves to pay closer attention in the future. The gHMBC cross peaks between the H3 signals at 2.76 & 3.16 ppm couple to the ^{13}C signal at 18 ppm, unfortuitously superimposed on the site 28 ^{13}C signal. We note that a gHMBC cross peak between the signals of the H3's and C28 is improbable.

We assign the other set of methylene NMR signals to site 22: $\delta_H = 3.69$ & 3.87 ppm and $\delta_C = 62.4$ ppm. We note that the H22 signals fall into a crowded portion of the 1-D ^{1}H NMR spectrum. Welcome to the NMR world of sugars.

One of the H3 signals, at 2.76 ppm, shares a gHMBC cross peak with a nonprotonated ^{13}C signal at 105.0 ppm. We assign this signal to site 10: $\delta_C = 105.0$ ppm. We note that oxygens at sites 5, 7, and 9 donate electron density to site 10 (as well as to sites 6 and 8) on the C5-C10 aromatic ring, thereby pushing the chemical shifts of the signals from sites 6, 8, and 10 upfield to lower values. The three electron-donating oxygens lower the C10 signal's chemical shift from near 128 ppm (benzene has a ^{13}C shift of 128.39 ppm) to 105 ppm.

We assign the last two downfield ^{1}H signals at 6.14 and 6.18 ppm to H6 and H8, but we cannot decide which is which. Sites 6 and 8 are chemically very similar, insofar as (1) their ^{1}H shifts are close and (2) they both couple equally to other nearby spins. If site 1 were a carbon instead of an oxygen, we might observe a gHMBC cross peak between the signals of H8 and C1, but it is not and so we do not. We resort to weaker arguments involving the electronegativity of atoms three bonds distant. The site 4 versus site 6 ^{13}C signals exhibit chemical shift differences that are fractionally greater (98.0 versus 96.9 ppm), so we focus on explaining these shift differences. The signal of C8 will be downfield relative to that of C6 because C2 is more electronegative than the site 5 hydroxyl deuteron (we expect the site 5 hydroxyl ^{1}H to have undergone chemical exchange with ^{2}H from the solvent). We assign site 8 as follows: $\delta_H = 6.14$ ppm and $\delta_C = 98.0$ ppm and for site 6 we write: $\delta_H = 6.18$ ppm and $\delta_C = 96.9$ ppm. A weak confirming argument can be made that H6, being closer to the deuteron on the site 5 hydroxyl group, may be slightly broader than H8, which is more bonds distant. We can see in the 1-D ^{1}H NMR spectrum that the more downfield ^{1}H resonance at 6.18 ppm is slightly shorter than its more upfield counterpart at 6.14 ppm. At least our two weak arguments (^{13}C shift of C8 signal is affected by C2 and H6 couples to the deuterium on the site 5 hydroxyl) are consistent with our assignment of site 6.

We now turn to the three nonprotonated ^{13}C resonances near 165 ppm. These three signals are associated with C5, C7, and C9. Examination of the downfield portion of the ^{13}C (f_1) chemical shift axis of the gHMBC spectrum reveals a cross peak between a midfield methine ^{1}H signal at 5.10 ppm and the ^{13}C signal at 166.7 ppm. The cross peak arises due to a coupling between H17 and C7 via an ether linkage. This shows us that the molecule likely has a single low-energy conformation which it adopts exclusively. If this were not the case, rotation about the C17-O-C7 ether linkage would prevent us from observing the gHMBC cross peak between the signals of H17 and C7. We assign site 7 as follows: $\delta_C = 166.7$ ppm and we assign site 17 as: $\delta_H = 5.10$ ppm and $\delta_C = 99.5$ ppm.

The C5 and C9 signals are, like the signals of the site 6 and 8 ^{1}H/^{13}C spin pairs, challenging to differentiate. We reason that the C9 signal is downfield from that of C5 based on C2 being more electronegative than the deuterium on the site 5 hydroxyl group. We also notice that it appears the gHMBC spectrum contains two relevant, overlapping cross peaks. The first of these two overlapping gHMBC cross peaks is observed between the H8 signal at 6.14 ppm and the ^{13}C signal at 165.1 ppm. The second of the overlapped cross peaks is between the H6 at 6.18 ppm and the ^{13}C signal at 164.8 ppm. Because H8 likely couples to C9 and H6 likely couples to C5, we assign the upfield ^{13}C signal to site 5 and the downfield signal to site 9. Given the coarse digital resolution of the gHMBC spectrum, it is reasonable to assert that our interpretation of the shape of the gHMBC signal intensity is at the limit of our confidence. We write for site 9: $\delta_C = 165.1$ ppm and for site 5: $\delta_C = 164.8$ ppm. We note in the 1-D ^{13}C NMR spectrum that the C9 signal is doubled while that from C5 is not. Given that C9 is much closer than C5 to site 2, which we believe is a racemic stereocenter we reason that having an *R*- versus and *S*-stereocenter at site 2 will affect site 9, two bonds distant, more than site 5, four bonds distant.

Having now exhausted all sugar substitutes, we turn our attention to assigning the only remaining ^{1}H and ^{13}C NMR signals of naringin, which all arise from the two cyclic sugars. Each cyclic sugar gives us two NMR entry points: the anomeric end of the spin system and the methylene/methyl end of the spin system. If we consider the top sugar ring as shown in Fig. 3.4.1, we observe that H17 is part of a ^{1}H spin system that extends through four methine groups to the site 22 methylene group. The spin system of the lower sugar ring runs from the anomeric H23 through four methine groups to the site 28 methyl group. The anomeric methine sites in cyclic sugars are easy to identify because the ^{1}H signals from them are the most downfield of all of the spins in the spin system. The anomeric carbon is bound to two oxygen atoms and its ^{13}C signal appears farthest downfield of the signals from the sugar ring, typically coming within 5 ppm of 100 ppm. We note that our C17 signal was found at 99.5 ppm. Newly outfitted with ^{13}C chemical shift information, we find the other anomeric ^{1}H/^{13}C pair in the lower midfield range of the HSQC spectrum very close to the site 17 cross peak. We write for site 23: $\delta_H = 5.24$ ppm and $\delta_C = 102.7$ ppm. Because we have already assigned all protonated sp^2-hybridized carbon sites, we are confident that this signal is attributable to our other anomeric ^{1}H/^{13}C spin pair.

Having assigned sites 17, 22, 23, and 28, we now delve into the unassigned midfield methine ^{1}H/^{13}C signal pairs, of which there are eight.

The H17 signal at 5.10 ppm shares a COSY cross peak with the midfield methine ^{1}H signal at 3.64 ppm. We assign this new methine signal to site 18: $\delta_H = 3.64$ ppm and $\delta_C = 79.3$ ppm. We confirm the site 18 assignment by noting that the H18 signal at 3.64 ppm participates in a gHMBC cross peak correlated with the C23 signal at 102.7 ppm. There is a $^3J_{CH}$ across the inter-ring C18-O-C23 ether linkage. This cross peak is consistent with our interpretation of the gHMBC cross peak between the signals of H17 and C7 as discussed previously.

The H23 signal at 5.24 ppm shares a COSY cross peak with the midfield methine ^{1}H signal at 3.93 ppm, to which we assign to site 24: $\delta_H = 3.93$ ppm and $\delta_C = 72.31$ ppm.

The H22 signals at 3.69 & 3.87 ppm share COSY cross peaks with the methine ^{1}H signal at 3.44 ppm, and so we assign the methine signal to site 21: $\delta_H = 3.44$ ppm and $\delta_C = 78.2$ ppm.

The H28 signal at 1.28 ppm also shares a gHMBC cross peak with the methine ^{13}C signal at 71.3 ppm. We assign this methine ^{1}H signal to site 27: $\delta_H = 3.88$ ppm and $\delta_C = 71.3$ ppm. We confirm the assignment, noting that the H28 signal at 1.28 ppm correlates in the COSY spectrum with a methine ^{1}H signal at 3.88 ppm. The H27 signal at 3.88 ppm is also observed to correlate in the gHMBC spectrum with the C28 signal at 18.4 ppm.

The last four methine sites are 19, 20, 25, and 26. Two pairs of the ^{1}H signals have similar chemical shifts: 3.38/3.39 ppm and 3.58/3.59 ppm. These two pairs of nearly isochronous (identical frequency) ^{1}H shifts require that we use the ^{13}C shifts to distinguish the methine spin pairs from one another.

The methyl H28 signal at 1.28 ppm shares a gHMBC cross peak with the unassigned methine ^{13}C signal at 74.0 ppm. Having already assigned C27, we attribute this unassigned ^{13}C signal to site 26: $\delta_H = 3.39$ ppm and $\delta_C = 74.0$ ppm.

The H18 signal at 3.64 ppm shares a gHMBC cross peak with the unassigned methine ^{13}C signal at 79.1 ppm. We assign this ^{13}C signal to site 19: $\delta_H = 3.58$ ppm and $\delta_C = 79.1$ ppm.

The H21 signal at 3.44 ppm correlates in the gHMBC spectrum to the methine ^{13}C signal at 71.3 ppm. We assign the 71.3 ppm signal to site 20: $\delta_H = 3.38$ ppm and $\delta_C = 71.3$ ppm.

The H24 signal at 3.93 ppm shares a gHMBC cross peak with the methine ^{13}C signal at 72.28 ppm, and so we assign site 25 as follows: $\delta_H = 3.59$ ppm and $\delta_C = 72.28$ ppm. The similarity of the site 24 ^{1}H and ^{13}C chemical shifts to others should arouse considerable skepticism as to the veracity of this assignment. We now must confirm as many of our last four assignments as possible. In essence, what we do is to account for every gHMBC cross peak. We can also look for notable absences, but this is less useful. For instance, if we have assigned sites 19 versus 25 correctly, we should *not* observe a gHMBC cross peak between the signals of H24 at 3.93 ppm and C19 at 79.1 ppm (we do not). We also fail to observe a gHMBC correlation between the signals of H18 at 3.64 ppm and C25 (or C24) at 72.3 ppm. These exercises often uncover erroneous assignments if these errors exist and also sharpen mental acuity. If we are conducting original research this activity is compulsory.

We observe a gHMBC cross peak between the H25 signal at 3.59 ppm and the C24 signal at 72.31, but we might worry that the cross peak is from H25 to the ^{13}C signal at 72.28 ppm. However, because C25 is the signal at 72.28 ppm, we can rule out this chemical shift as the possible source of the ^{13}C spin involved in the generation of this particular gHMBC cross peak, because the H25 signal will not correlate with the C25 signal in the gHMBC spectrum because the spins are separated by only one bond (both are from site 25). Our H27 signal at 3.88 ppm shares a confirming gHMBC cross peak to the C26 signal at 74.0 ppm.

The overlap found for the methine ^{1}H signals of the sugar rings in naringin limits the usefulness of the COSY spectrum and forces us to rely on information culled from the gHMBC spectrum. Sometimes the other spectra serve only to ready us for understanding the gHMBC spectrum. That is, the 1-D ^{1}H and ^{13}C spectra, combined with the 2-D COSY and HSQC spectra, provide the foundation upon which the edifice of site-to-signal assignment of complex molecules is built using the gHMBC spectrum.

3.5 (+)-RUTIN HYDRATE IN METHANOL-D$_4$

The reader may note a similarity between this molecule and the previous one. Rutin is a glycoside with antibacterial properties made by the ice plant (*Carpobrotus edulis*). Rutin is also found especially in the rinds of citrus fruits. Rutin also has 27 carbons like naringin, but has fewer hydrogens and so assigning some of its nonprotonated ^{13}C resonances is more challenging. Rutin is derived from coupling the flavonol quercetin and the disaccharide rutinose. Of rutin's 27 carbons, 15 are methines and 10 are nonprotonated. This leaves one methyl and one methylene which are therefore trivial to assign. Rutin also boasts 16 oxygen atoms, or more than one for every other carbon atom—we observe many downfield NMR resonances. With two aromatic rings, a heterocyclic ring with a ketone carbonyl conjugated to a gamma oxygen, and two sugar rings, rutin has an impressive IHD of 13, giving an empirical formula of $C_{27}H_{30}O_{16}$. The structure of (+)-rutin is found in Fig. 3.5.1. Rutin's 16 oxygen atoms are found in six hydroxyl groups on its two sugar rings, four phenolic hydroxyls, five ether linkages, including the one in the heterocycle, and on the carbonyl carbon at site 4.

There is no c_2 symmetry axis for the aromatic ring at the top-right of the shown structure as the site 13 hydroxyl group disrupts the symmetry found in the analogous region of the previous molecule (naringin). We observe as many ^{13}C signals as there are carbon sites in the molecule.

A sample of (+)-rutine hydrate was dissolved in methanol-d_4. This sample was then used to collect the 1-D ^{1}H NMR spectra shown in Figs. 3.5.2 and 3.5.3, as well as the 1-D ^{13}C NMR spectrum shown in Fig. 3.5.4. Three 2-D spectra were also collected using the sample. The 2-D ^{1}H-^{1}H COSY NMR spectrum is shown in Fig. 3.5.5. An expanded portion of the COSY spectrum appears in Fig. 3.5.6. The 2-D ^{1}H-^{13}C HSQC NMR spectrum is appears in Fig. 3.5.7. An expanded portion of the HSQC spectrum is found in Fig. 3.5.8. The 2-D ^{1}H-^{13}C gHMBC NMR spectrum can be found in Fig. 3.5.9. An expansion of the midfield region of the gHMBC spectrum is shown in Fig. 3.5.10. An expansion of the downfield region of the gHMBC spectrum is found in Fig. 3.5.11.

We begin by predicting the multiplicities we expect to observe, and this time we also take the time to predict when we expect to observe signal splittings from the $^4J_{HH}$'s between H12 and H16, as well as from the coupling between H6 and H8. We put the predicted multiplicities of rutin into Table 3.5.1. Into Table 3.5.2 we place our accounting of the groups found at the numbered molecular sites for rutin. Most importantly, we place into Table 3.5.3 our careful tabulation of the shifts we observe in the 1-D ^{13}C and 2-D ^{1}H-^{13}C HSQC NMR spectra. While it is true that the chemical shift should be reported as the midpoint of the integral of the 1-D ^{1}H NMR spectral resonance for a given ^{1}H, interpretation of a 2-D spectrum, be it a COSY, HSQC, or gHMBC, is made easier by measuring and employing chemical shifts that line up with the middle of the entire ^{1}H resonance. For strongly nonfirst-order multiplets, using the actual chemical shift will make interpretation of the 2-D spectra significantly more challenging, insofar as it will cause us to look for cross peaks at locations slightly different from their actual location. There are compelling arguments on both sides of the integral-center versus cross-peak-center debate. We are assigning 2-D spectra here and so we will use cross-peak-center chemical shift values.

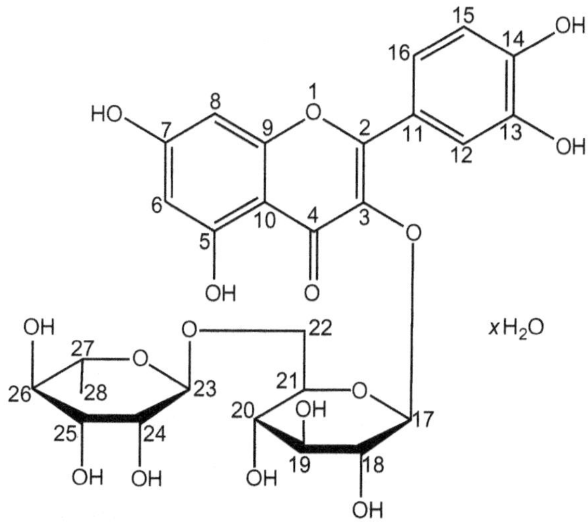

FIG. 3.5.1 The structure of rutin hydrate.

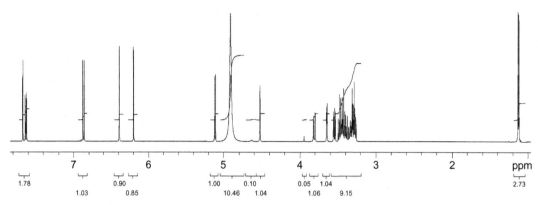

FIG. 3.5.2   The 1-D ^{1}H NMR spectrum of rutin hydrate in methanol-d_4.

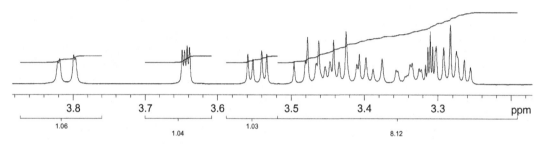

FIG. 3.5.3   An expanded portion of the 1-D ^{1}H NMR spectrum of rutin hydrate in methanol-d_4.

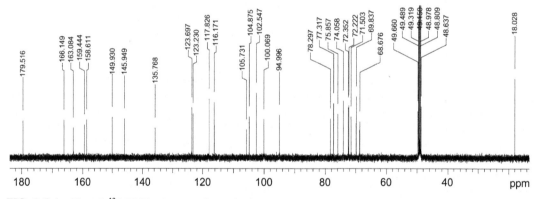

FIG. 3.5.4   The 1-D ^{13}C NMR spectrum of rutin hydrate in methanol-d_4.

The unique chemical shift of the carbonyl ^{13}C resonance is our starting point. We assign the site 4 signal as follows: $\delta_C = 179.5\,\text{ppm}$. We might also assign the only methylene and methyl signals, but we defer those assignments for the sake of continuity.

Of the five aromatic ^{1}H signals, we expect those of H6 and H8 will appear upfield from those of H12, H15, and H16. Sites 6 and 8 are both *ortho* or *para* to three electron donating groups,

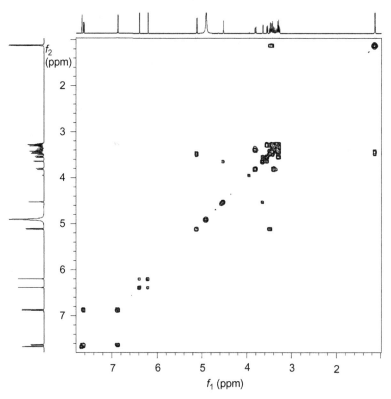

FIG. 3.5.5 The 2-D ^{1}H-^{1}H COSY NMR spectrum of rutin hydrate in methanol-d_4.

while sites 12, 15, and 16 only receive electron density donation from a single electron donating group that is either *ortho* or *para*. That is, upfield shift contributions are additive for sites 6 and 8, but not for sites 12, 15, and 16. The ^{1}H signals from the spins at sites 15 and 16 are expected to share a strong COSY cross peak in the downfield region. We observe a strong COSY correlation between aromatic ^{1}H signals at 6.87 and 7.63 ppm. We look in the 1-D ^{1}H NMR spectrum to ensure that both of these aforementioned resonances feature the large *cis*-$^3J_{HH}$ we expect between H15 and H16 of about 8 Hz. We assign to site 16 the ^{1}H resonance that is shorter in height because the H16 resonance is additionally split via a $^4J_{HH}$ by H12. The most downfield resonance at 7.63 ppm features the d^2 multiplicity pattern and we assign it to site 16 as follows: $\delta_H = 7.63$ ppm and $\delta_C = 123.7$ ppm. We therefore also assign site 15: $\delta_H = 6.87$ ppm and $\delta_C = 116.2$ ppm.

If we look more closely at the COSY spectrum, we can pick out another pair of cross peaks involving the H16 signal. We observe a correlation between the H16 signal at 7.67 ppm and a ^{1}H signal at 7.63 ppm that is easy to miss, the correlation being found very close to intensity on the diagonal of the COSY spectrum. We assign this downfield ^{1}H signal at 7.67 ppm to site 12: $\delta_H = 7.67$ ppm and $\delta_C = 117.8$ ppm.

The ^{1}H signals of H12 at 7.67 ppm, of H16 at 7.63 ppm, and of H15 at 6.87 ppm share gHMBC cross peaks with the nonprotonated ^{13}C signal at 149.9 ppm. We assign this ^{13}C signal to site 14: $\delta_C = 149.9$ ppm, noting that H16 would be unlikely to couple to C13 four bonds away, therefore ruling out site 13 for consideration for assignment to this ^{13}C signal.

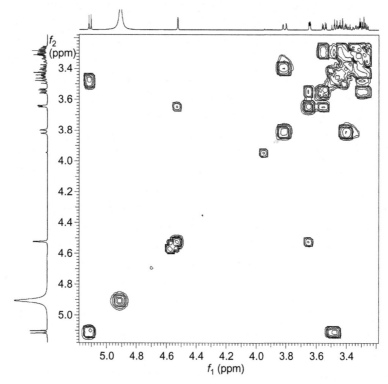

FIG. 3.5.6   An expanded portion of the 2-D ^{1}H-^{1}H COSY NMR spectrum of rutin hydrate in methanol-d_4.

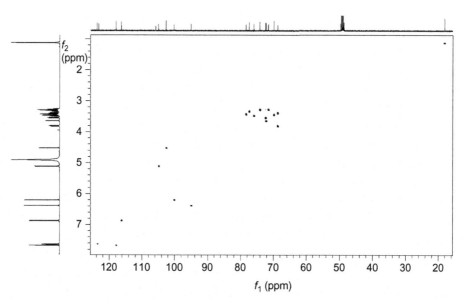

FIG. 3.5.7   The 2-D ^{1}H-^{13}C HSQC NMR spectrum of rutin hydrate in methanol-d_4.

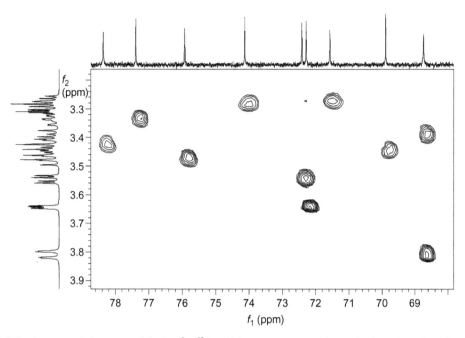

FIG. 3.5.8 An expanded portion of the 2-D ^{1}H-^{13}C HSQC NMR spectrum of rutin hydrate in methanol-d_4.

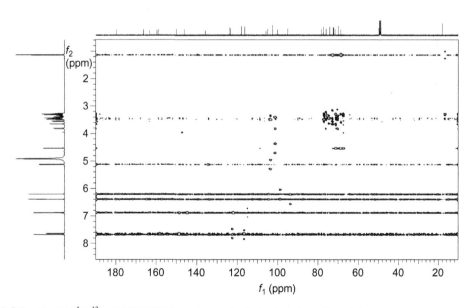

FIG. 3.5.9 The 2-D ^{1}H-^{13}C gHMBC NMR spectrum of rutin hydrate in methanol-d_4.

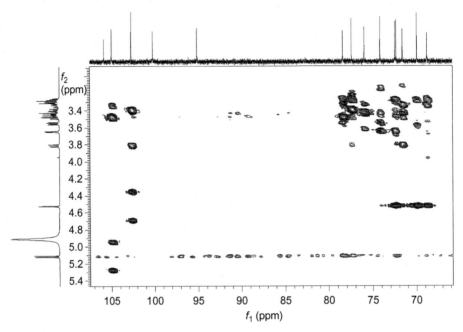

FIG. 3.5.10   The midfield portion of the 2-D ^{1}H-^{13}C gHMBC NMR spectrum of rutin hydrate in methanol-d_4, expanded for clarity.

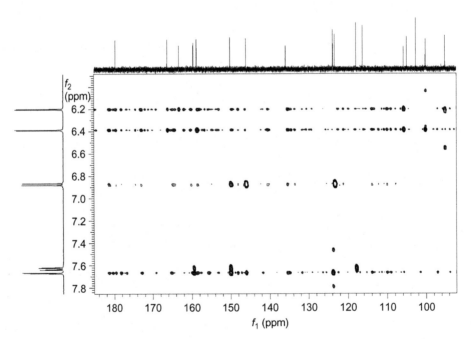

FIG. 3.5.11   The downfield portion of the 2-D ^{1}H-^{13}C gHMBC NMR spectrum of rutin hydrate in methanol-d_4, expanded for clarity.

TABLE 3.5.1 Predicted Multiplicities for the ^{1}H's of Rutin

Site in Molecule	Expected Multiplicity
6	d^a
8	d^a
12	d^a
15	d
16	d^{2a}
17	d
18	d^2
19	d^2
20	d^2
21	d^3
22	2×d^2
23	d
24	d^2
25	d^2
26	d^2
27	dq
28	d

^{a}Includes $^4J_{HH}$.

TABLE 3.5.2 Numbered Molecular Sites of Rutin

Group Type	Site Number
CH$_3$ (methyl)	28
CH$_2$ (methylene)	22
CH (methine)	6, 8, 12, 15, 16, 17, 18, 19, 20, 21, 23, 24, 25, 26, 27
C$_{np}$ (nonprotonated)	2, 3, 4, 5, 7, 9, 10, 11, 13, 14
numbered heteroatom	1 (O)

The H12 signal at 7.67 ppm shares a weak gHMBC cross peak with a nonprotonated ^{13}C signal at 145.9 ppm. We assign this signal to site 13: $\delta_C = 145.9$ ppm. Notice that we do not observe a gHMBC cross peak between the signals of H16 at 7.63 ppm and C13 at 145.9 ppm.

We have yet to assign the ^{1}H and ^{13}C NMR signals of the aromatic ring involving C5-C10. None of the ^{13}C signals from the aromatic ring involving sites 5–10 will be found within

TABLE 3.5.3 ^{1}H and ^{13}C NMR Signals of Rutin Listed by Group Type

^{1}H Signal (ppm)	^{13}C Signal (ppm)	Group Type
–	179.5	Nonprotonated
–	166.1	Nonprotonated
–	163.1	Nonprotonated
–	159.4	Nonprotonated
–	158.6	Nonprotonated
–	149.9	Nonprotonated
–	145.9	Nonprotonated
–	135.8	Nonprotonated
7.63	123.7	Methine
–	123.2	Nonprotonated
7.67	117.8	Methine
6.87	116.2	Methine
–	105.7	Nonprotonated
5.11	104.9	Methine
4.53	102.5	Methine
6.19	100.1	Methine
6.38	95.0	Methine
3.43	78.3	Methine
3.33	77.3	Methine
3.48	75.9	Methine
3.28	74.1	Methine
3.54	72.4	Methine
3.64	72.2	Methine
3.27	71.5	Methine
3.44	69.8	Methine
3.39 & 3.81	68.7	Methylene
1.13	18.0	Methyl

10 ppm of benzene (128.39 ppm). Because of the 1,3,5-substitution pattern of electron donating groups (oxygens) on this aromatic ring, the protonated sites will experience upfield shifts near 100 ppm (C6, C8, C10), while those nonprotonated ^{13}C's directly bound to oxygen will, through the direct pull of the highly electronegative oxygen atoms, generate downfield ^{13}C signals near 150 ppm (C5, C7, C9). This helps us eliminate some of the downfield ^{13}C signals

from consideration for the 1,2-substituted C11-C16 aromatic ring. We expect the H15 signal to share a strong cross peak with that of C11 in the gHMBC spectrum, because of the *trans*-$^3J_{CH}$ between the two spins. We observe a strong gHMBC cross peak between the H15 signal at 6.87 ppm and the nonprotonated ^{13}C signal at 123.2 ppm. We assign the ^{13}C signal to that of site 11: $\delta_C = 123.2$ ppm.

Just as we used the *trans*-$^3J_{CH}$ to correlate the signals of H15 and C11, so too do we use the signals of H16 and H12 to correlate with the C2 signal. The H16 signal at 7.63 ppm and the H12 signal at 7.67 ppm correlate with a downfield nonprotonated ^{13}C signal at 159.4 ppm in the gHMBC spectrum. This ^{13}C signal we assign to site 2: $\delta_C = 159.4$ ppm. We expected that the ^{13}C signals of sites 2 and 3 would be found near 150 ppm, but the carbonyl group at site 4 has in hindsight moved the C3 signal downfield by 10 ppm, so perhaps the C2 signal has been moved up by 10 ppm to maintain balance through alternation, which is often found to be the case. The H12 signal at 7.67 ppm shares a moderate intensity gHMBC cross peak with the nonprotonated aromatic ^{13}C resonance at 135.8 ppm. This interaction must arise from a $^4J_{CH}$ between H12 and C3. We write for site 3: $\delta_C = 135.8$ ppm. Gratifyingly, this ^{13}C resonance correlates in the gHMBC spectrum with one of the two anomeric ^{1}H signals at 5.11 ppm from the sugar portion of rutin.

We now assign the NMR signals of the C5-C10 ring. The H6 and H8 signals can be identified as a pair because they are the only two downfield ^{1}H resonances yet to be assigned (there were only five to begin with). These ^{1}H signals are found at 6.19 and 6.38 ppm, where we would expect to find ^{1}H signals on an aromatic ring with electron donating groups. H6 and H8 both correlate with the nonprotonated upfield aromatic ^{13}C signal at 105.7 ppm. We assign this signal to site 10: $\delta_C = 105.7$ ppm. Electron donation by lone pairs on the oxygens at sites 5, 7, and 9 serve to move the C10 resonance upfield from the benzene expectation value of 128.39 ppm.

Examination of the structure of rutin allows us to recognize the similar chemical environment shared between the three spins of sites 5 + 6 and the three spins of sites 9 + 8 (C5's chemical environment is a lot like C9's, C6's environment is like that of C8's, H6's environment is like that of H8's). The ^{1}H shift differences for the signals of H6 versus H8 are about 0.2 ppm and the ^{13}C shift differences between the signals of C6 and C8 are 4 ppm, and so we must be cautious in the application of reasoning based on chemical shifts to differentiate between the sites. We have yet to pair C5, C7, and C9 to the three unassigned ^{13}C signals downfield at 166.1, 163.1, and 158.6 ppm. The ^{1}H signal from H6 or H8 at 6.38 ppm shares a strong gHMBC cross peak with the downfield nonprotonated ^{13}C signal at 158.6 ppm. This cross peak is assigned to the H8-C9 interaction. If only one cross peak is going to be observed, it likely corresponds to this unique site (put another way, if the cross peak correlates the signals of H6 and C7, why do we not observe a similar correlation between the signals of H6 and C5, or between the signals of H8 and C7?). The assignment of this one gHMBC cross peak could be in error, but we have no way to know better other than using chemical shift prediction. Note that H6 and H8 are both geometrically aligned to generate similar $^4J_{CH}$'s in the shape of a W to C4.

With little recourse, we venture into the chemical shift argument. H6 will receive more donated electron density from the *ortho* hydroxyl group at site 5 than H8 will receive from the *ortho* oxygen at site 1. The hydroxyl at site 7 should donate electron density equally to H6 and H8. The *para* resonance contributions are typically smaller, so we neglect these for the sake of brevity. The site 4 carbonyl group withdraws electron density from the adjacent oxygen at site 1 and attached to C9 through conjugation. The pull of electron density from the site 1 oxygen atom prevents this oxygen from donating its electron density to the aromatic ring. This distracting pull from the carbonyl causes site 8 to receive less donated electron density and therefore site 8 is less shielded, meaning that the H8 signal is shifted downfield to a higher chemical shift. We write for site 8: $\delta_H = 6.38$ ppm and $\delta_C = 95.0$ ppm and for site 9: $\delta_C = 158.6$ ppm.

Because we have assigned one pair of $^1H/^{13}C$ signals to the site 8 spins, we assign the other pair to site 6: $\delta_H = 6.19$ ppm and $\delta_C = 100.1$ ppm.

The H6 signal at 6.19 ppm and the H8 signal at 6.38 ppm share gHMBC cross peaks with the nonprotonated ^{13}C resonance at 166.1 ppm, while only the H6 signal at 6.19 ppm correlates with the other downfield aromatic nonprotonated ^{13}C resonance at 163.1 ppm. Because C7 is between H6 and H8, C7 is expect to couple to both H6 and H8, while C5 is not expected to couple to both H6 and H8. We assign the last two downfield nonprotonated ^{13}C signals with the reasoning the C7 will couple equally to both aromatic ^1H's while C5 will not. We write for site 7: $\delta_C = 166.1$ and for site 5: $\delta_C = 163.1$ ppm.

Looking back at these rather tenuous assignments, we can take some small solace in noting that the C5 and C7 signals have chemical shifts that are closer to one another than to that of C9. The chemical environments of C5 and C7 are more similar to one another's than to that of C9. If the H6 and H8 assignments are reversed, we would assign the 163.1 ppm ^{13}C signal to C9. Given ample instrument time, this question could be resolved, but the nature of applied NMR spectroscopy sometimes is such that we will be unable to arrive at answers with certainty. It is important to identify portions of assignments where we are less confident as to veracity.

We now address the two cyclic sugar rings of rutin at numbered sites 17–28 in Fig. 3.5.1. We start by identifying the two anomeric sites, the methylene spin system terminus, and the methyl spin system terminus. The anomeric methine ^1H NMR signal at 5.11 ppm correlates with the C3 signal at 135.8 ppm in the gHMBC spectrum. We assign this methine ^1H signal to site 17: $\delta_H = 5.11$ ppm and $\delta_C = 104.9$ ppm.

We identify the other anomeric $^1H/^{13}C$ NMR signal pair with a ^{13}C chemical shift near 100 ppm in the midfield region of the HSQC spectrum as that belonging to site 23: $\delta_H = 4.53$ ppm and $\delta_C = 102.5$ ppm.

We assign the methyl $^1H/^{13}C$ signals from site 28 as follows: $\delta_H = 1.13$ ppm and $\delta_C = 18.0$ ppm. Using the HSQC spectrum (or the second to last row in Table 3.5.3, which was derived principally from the HSQC spectrum), we assign site 22 as follows: $\delta_H = 3.39$ & 3. 81 ppm and $\delta_C = 68.7$ ppm.

What remains is for us to assign the NMR signals from eight methine groups that all experience similar chemical environments. We use our assignment methodology to work from the known ends of each spin systems to the unknown spins the middle of the sugar rings. We begin by using the COSY spectrum, hopefully using the H17 signal to find the H18 signal, the H22 signals to find the H21 signal, the H23 signal to find the H24 signal, and the H28 signal to find the H27 signal.

In the COSY spectrum, the H28 signal at 1.13 ppm correlates with a ^1H signal at 3.43 or 3.44 ppm. We must now use the gHMBC to differentiate between the two $^1H/^{13}C$ spin pairs whose ^1H shifts are within 0.01 ppm of one another. The H28 signal at 1.13 ppm also shares a gHMBC correlation with the methine ^{13}C signal at 69.8 ppm, but not 78.3 ppm. We assign the 3.44 ppm signal to site 27: $\delta_H = 3.44$ ppm and $\delta_C = 69.8$ ppm.

The H23 signal at 4.53 ppm shares a COSY cross peak with the resolved ^1H signal at 3.64 ppm. We assign the 3.64 ppm ^1H signal to site 24: $\delta_H = 3.64$ ppm and $\delta_C = 72.2$ ppm. We confirm our assignment in the gHMBC by noting that the H23 signal at 4.53 ppm correlates with the C24 signal at 72.2 ppm. There is no correlation observed between the H24 signal at 3.64 ppm and the C23 signal at 102.5 ppm. The C23 signal at 102.5 ppm is found to correlate with the H22 signals at 3.39 & 3.81 ppm. We note that at the very least the C23 signal did not correlate with some other methine ^{1}H signal.

The H24 signal is expected to share a pair of cross peaks with the H25 signal in the COSY spectrum. The H24 signal at 3.64 ppm shares a COSY cross peak with the ^{1}H signal at 3.54 ppm. We assign the 3.54 ppm ^{1}H signal to site 25: $\delta_H = 3.54$ ppm and $\delta_C = 72.4$ ppm. We confirm our assignment, noting that the H24 signal at 3.64 ppm shares a gHMBC cross peak with the ^{13}C signal at 72.4 ppm (the C24 signal is observed at 72.2 ppm, so we are confident that the gHMBC cross peak we observe is not attributable to an H24-C24 interaction because it would be split along the f_2 axis by the $^1J_{CH}$).

The H25 signal at 3.54 ppm shares a COSY cross peak with the ^{1}H signal at 3.27 or 3.28 ppm. The H25 signal is also found to correlate with the ^{13}C signal at 74.1 ppm but not with that at 71.5 ppm, allowing us to choose the correct ^{1}H/^{13}C spin pair for site 26: $\delta_H = 3.28$ ppm and $\delta_C = 74.1$ ppm.

Completing our assignment of the spins from sites 23 to 28, we look for the COSY correlation between the signals of H26 and H27, and the gHMBC correlations between the signals of H26 and C27 as well as between the signals H27 and C26. We observe a COSY cross peak between the H26 signal at 3.28 ppm and that of H27 at 3.44 ppm, but both of these shifts are close to other observed ^{1}H resonances so this cross peak does not unambiguously confirms our assignment of site 26. For additional confirmation, we observe that the H26 signal at 3.28 ppm shares a gHMBC cross peak with the C27 signal at 69.8 ppm and also that the H27 signal at 3.44 ppm shares a gHMBC cross peak with the C26 signal at 74.1 ppm.

Looking to the other ring, we start at the anomeric site 17. The H17 signal at 5.11 ppm shares a COSY cross peak with the methine ^{1}H signal at 3.48 ppm. This 3.48 ppm ^{1}H signal is assigned to site 18: $\delta_H = 3.48$ ppm and $\delta_C = 75.9$ ppm. We confirm that the H18 signal at 3.48 ppm participates in a confirming gHMBC cross peak with the C17 signal at 104.9 ppm. Interestingly, the H17 signal at 5.11 ppm fails to correlate with the C18 signal at 75.9 ppm in the gHMBC spectrum. The C18 signal at 75.9 ppm only shares a gHMBC cross peak with one methine ^{1}H signal, and it appears that the H17 signal is not the one.

The H18 signal at 3.48 ppm shares a pair of COSY cross peaks with a nearby ^{1}H signal at 3.43 or 3.44 ppm. Because we have already assigned the 3.44 ppm ^{1}H signal to H27, we assign the 3.43 ppm signal to site 19: $\delta_H = 3.43$ ppm and $\delta_C = 78.3$ ppm. We observe the two gHMBC cross peaks that confirm the site 19 assignment: we observe a cross peak between the H19 signal at 3.43 ppm and the C18 signal at 75.9 ppm and also between the H18 signal at 3.48 ppm and the C19 signal at 78.3 ppm.

The site 20 and site 21 methines are now the last two ^{1}H/^{13}C spin pairs to be assigned. One of the unassigned methine signals (at 3.33 ppm) shares a gHMBC correlation with the C17 signal at 104.9 ppm. This ^{1}H signal is assigned to site 21: $\delta_H = 3.33$ ppm and $\delta_C = 77.3$ ppm. By the process of elimination we assign the last pair of ^{1}H/^{13}C signals to site 20: $\delta_H = 3.27$ ppm and $\delta_C = 71.5$ ppm. The H21 signal at 3.33 ppm is found to correlate in the gHMBC spectrum with the signals of C19 at 78.3 ppm, C18 at 75.9 ppm, C20 at 71.5 ppm, and C22 at 68.7 ppm. The H22 signals at 3.39 & 3.81 ppm correlate with the C21 signal at 77.3 ppm in the gHMBC spectrum, but perhaps more weakly than we expect. The H20 signal at 3.27 ppm is found to correlate in the gHMBC spectrum with the C21 signal at 77.3 ppm and with the C19 signal at 78.3 ppm. Finally, the H19 signal at 3.43 ppm is observed to share a gHMBC cross peak with the C20 signal at 71.5 ppm. The cross peaks between the signals of H19 and H20 are unfortunately found in a crowded region of the COSY spectrum.

Two Twenty-Carbon Natural Products, One With an Unknown Structure

In 2014, a Chemical Engineering student named Steven Edgar approached the author with an interesting NMR project. It started with the assignment of NMR signals to a known structure (taxadiene), i.e., as an assignment problem, but immediately thereafter data was being collected on a related C_{20} unknown thought to be generated via an epoxide intermediate. Instructors are rarely faced with true unknowns, because the instructor normally chooses interesting molecules of known structure to render as an unknown problem. Most often, a researcher presents a molecule of known structure for NMR analysis.

In this case, however, only the empirical formula was known. The NMR spectra were used to determine how many methyl, methylene, methine, and nonprotonated carbons were present. Only in mechanistic retrospect was it possible to fully grasp how the new molecule was derived from taxadiene. In particular, a tertiary carbocation rearrangement was found to have occurred—an unanticipated mechanistic step.

Resonance overlap in the ^1H chemical shift range was present and this required entertaining multiple possibilities in parallel. The worst overlap was present between three pairs of ^1H signals at 1.25–1.27 ppm, at 1.55–1.56 ppm, and at 1.71–1.72 ppm, meaning that eight (2^3) distinct interpretations were possible for these six ^1H signals. In the ^{13}C dimension, sufficient chemical shift dispersion allowed the unambiguous measurement when interpreting the 2-D ^1H-^{13}C gHMBC NMR spectrum except for two ^{13}C signals at 34.0 and 34.1 ppm which were sometimes challenging to distinguish from one another in the 2-D gHMBC spectrum with its coarser digital resolution.

4.1 TAXADIENE IN BENZENE-D_6

The taxadiene assignment is relatively straightforward given that its structure is known. Fig. 4.1.1 shows the structure of taxadiene. Taxadiene has an empirical formula of $C_{20}H_{32}$. With an index of hydrogen deficiency of five, taxadiene possesses three rings and two carbon-carbon double bonds. By examining the structure of taxadiene, we generate two tables. Our first table,

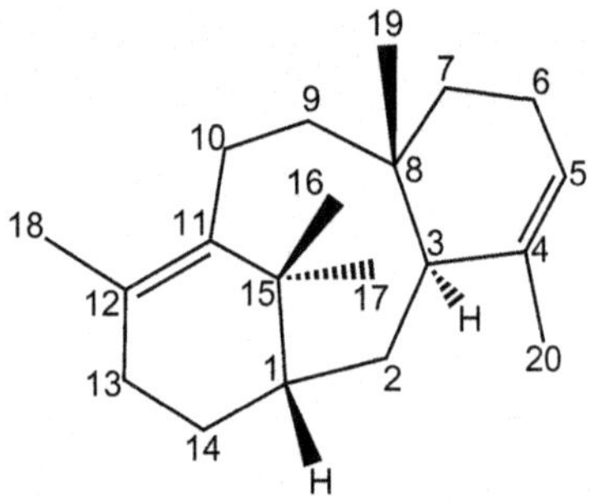

FIG. 4.1.1   The structure of taxadiene.

Table 4.1.1, contains the resonance multiplicities we predict for the ^{1}H signals of taxadiene. Table 4.1.2 lists the numbered sites of taxadiene by group type.

A small amount of taxadiene was dissolved in benzene-d_6 and used to obtain the NMR spectra in this section. Fig. 4.1.2 shows the 1-D ^{1}H NMR spectrum of taxadiene in benzene-d_6 and Fig. 4.1.3 shows taxadiene's 1-D ^{13}C NMR spectrum. The 2-D ^{1}H-^{1}H COSY NMR spectrum of taxadiene appears in Fig. 4.1.4. An expanded portion of the COSY spectrum of taxadiene is displayed in Fig. 4.1.5. The 2-D ^{1}H-^{13}C HSQC NMR spectrum of taxadiene is found in Fig. 4.1.6,

TABLE 4.1.1   Predicted Multiplicities for the ^{1}H Resonances of Taxadiene

Site in Molecule	Expected Multiplicity
1	d^4
2	$2 \times d^3$
3	d^2
5	d^2
6	$2 \times d^4$
7	$2 \times d^3$
9	$2 \times d^3$
10	$2 \times d^3$
13	$2 \times d^3$
14	$2 \times d^4$
16	s
17	s
18	s
19	s
20	s

TABLE 4.1.2   Numbered Sites of Taxadiene by Group Type

Group Type	Site Number
CH$_3$ (methyl)	16, 17, 18, 19, 20
CH$_2$ (methylene)	2, 6, 7, 9, 10, 13, 14
CH (methine)	1, 3, 5
C$_{np}$ (nonprotonated)	4, 8, 11, 12, 15

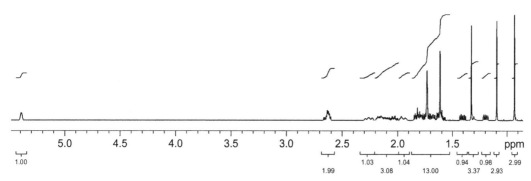

FIG. 4.1.2   The 1-D ^{1}H NMR spectrum of taxadiene in benzene-d_6.

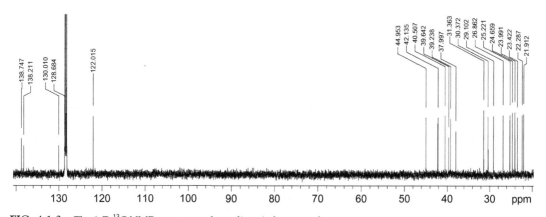

FIG. 4.1.3   The 1-D ^{13}C NMR spectrum of taxadiene in benzene-d_6.

with an expanded portion in Fig. 4.1.7. The 2-D ^{1}H-^{13}C gHMBC NMR spectrum of taxadiene is shown in Fig. 4.1.8, with two additional expansions in Figs. 4.1.9 and 4.1.10. A 2-D ^{1}H-^{1}H NOESY NMR spectrum of the taxadiene sample obtained using a mixing time of 350 ms appears in Fig. 4.1.11. An expansion of the same NOESY spectrum is shown in Fig. 4.1.12.

There are several entry points we may use: the methyl groups (16–20), the four sp^2-hybridized carbons (4, 5, 11, 12), but more useful is the uniquely downfield H5 signal,

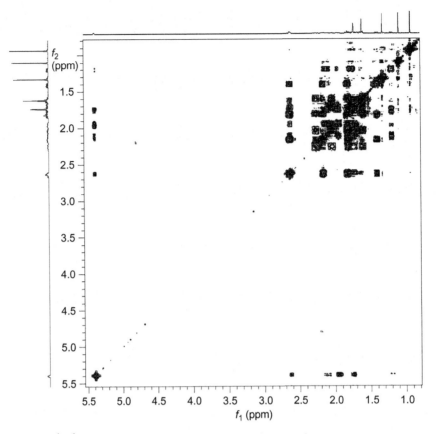

FIG. 4.1.4   The 2-D ^{1}H-^{1}H COSY NMR spectrum of taxadiene in benzene-d_6.

because H5 is the only proton bound to an sp^2-hybridized carbon. The H5 signal is observed at 5.39 ppm. In the HSQC spectrum, we find the cross peak in which the C5 resonance participates at 122.0 ppm. We write for site 5: $\delta_H = 5.39$ ppm and $\delta_C = 122.0$ ppm. We could have identified the C5 signal just from the 1-D ^{13}C spectrum: Because C5 is the only protonated sp^2-hybridized ^{13}C, it relaxes more quickly than its nonprotonated counterparts and therefore generates a more intense signal in the 1-D ^{13}C spectrum than its shift-similar, nonprotonated counterparts. The protonated ^{13}C NMR signals are typically more intense because the 1-D ^{13}C spectrum was acquired with "typical" relaxation delay parameter setting. That is, the delay between successive scans was less than the two or three times the T_1 relaxation times of the nonprotonated sp^2-hybridized carbons in taxadiene.

Before we proceed further, we examine the HSQC spectrum and group the ^{1}H and ^{13}C signals by site, placing this information into Table 4.1.3. As we proceed, we must ensure that the integrated intensity of the ^{1}H signals in the 1-D ^{1}H NMR spectrum (Fig. 4.1.2) accurately reflects the number of ^{1}H sites we list in the various rows of Table 4.1.3. We must list in our table the correct number of methyls (5), methylenes (7), and methines (3). In addition, we also list the four nonprotonated ^{13}C signals that we observe in the 1-D ^{13}C NMR spectrum

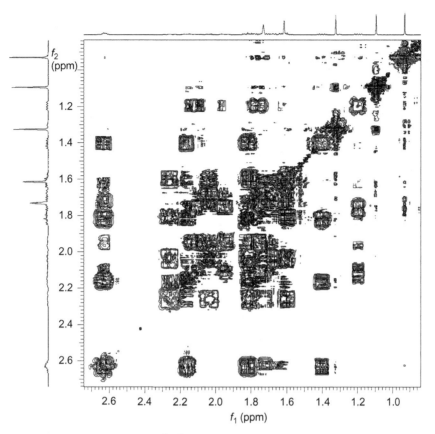

FIG. 4.1.5 An expanded portion of the 2-D ^{1}H-^{1}H COSY NMR spectrum of taxadiene in benzene-d_6.

(Fig. 4.1.3). We know that a ^{13}C signal corresponds to a nonprotonated site if the signal does not participate in any observed HSQC cross peaks (barring the occasional exception, as from an unusual $^1J_{CH}$, for example).

As we proceed with the assignment, we correlate pairs of methylene ^{1}H signals and individual methyl and methine ^{1}H signals to rows in Table 4.1.3. We also use our knowledge of how chemical environment affects ^{13}C chemical shifts and how long-range J_{CH} couplings generate gHMBC cross peaks to assign the nonprotonated ^{13}C signals. Table 4.1.3 is populated by moving left to right along the f_1 (^{13}C) chemical shift axis of the HSQC spectrum. Entries for nonprotonated ^{13}C's are included for completeness. These entries are known by noting in the 1-D ^{13}C spectrum across the top of the HSQC figure the signals for which there is no corresponding HSQC cross peak.

In the 2-D ^{1}H-^{1}H COSY NMR spectrum of taxadiene, the H5 signal at 5.39 ppm participates in three pairs of cross peaks. The H5 signal shares a weak cross peak with a methine or methylene ^{1}H signal at 2.63 ppm. H5 also shares a strong cross peak with a methylene ^{1}H signal at 1.95 ppm and shares a medium-to-strong cross peak with either a methylene ^{1}H signal at 1.73 ppm or to a methyl ^{1}H signal at 1.72 ppm. The narrowness of the cross peak between H5

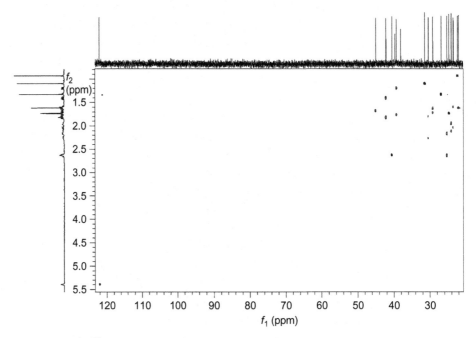

FIG. 4.1.6   The 2-D ^{1}H-^{13}C HSQC NMR spectrum of taxadiene in benzene-d_6.

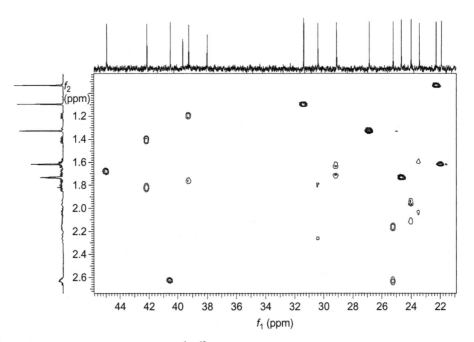

FIG. 4.1.7   An expanded portion of the 2-D ^{1}H-^{13}C HSQC NMR spectrum of taxadiene in benzene-d_6.

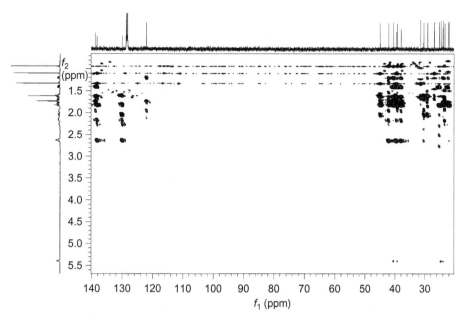

FIG. 4.1.8   The 2-D ^{1}H-^{13}C gHMBC NMR spectrum of taxadiene in benzene-d_6.

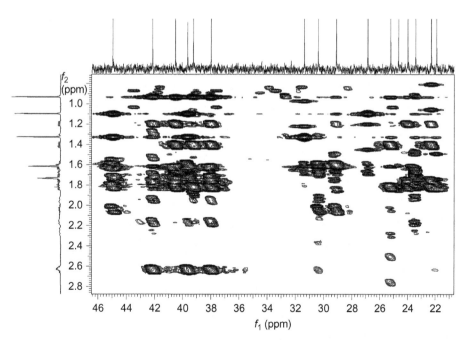

FIG. 4.1.9   The upfield portion of the 2-D ^{1}H-^{13}C gHMBC NMR spectrum of taxadiene in benzene-d_6, expanded for clarity.

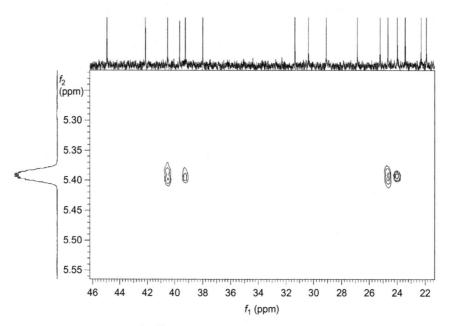

FIG. 4.1.10 The portion of the 2-D ¹H-¹³C gHMBC NMR spectrum of taxadiene in benzene-d_6 showing cross peaks involving the most downfield ¹H NMR resonance of taxadiene.

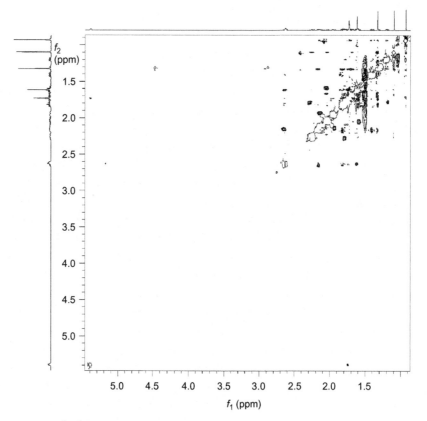

FIG. 4.1.11 The 2-D ¹H-¹H NOESY NMR spectrum of taxadiene in benzene-d_6 obtained with a mixing time of 350 ms.

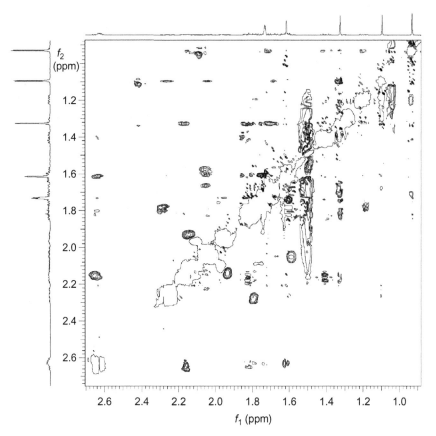

FIG. 4.1.12 An expansion showing the upfield portion the 2-D ^{1}H-^{1}H NOESY NMR spectrum of taxadiene in benzene-d_6 obtained with a mixing time of 350 ms.

and the ^{1}H signal at 1.72–1.73 ppm suggests that the resonance to which H5 is coupled in this region is not strongly coupled to other spins. This in turn suggests that the resonance at 1.72–1.73 ppm to which H5 is coupled is not that of H6 or H7 as all of these signals should be broadened from extensive coupling, but rather from the isolated site 20 methyl ^{1}H's. We write for site 20: $\delta_H = 1.73$ ppm and $\delta_C = 24.7$ ppm.

If we examine the taxadiene structure, we can reasonably assume that H5 will couple to at least one of the site 6 methylene ^{1}H's, as well as to the site 20 methyl ^{1}H's, and possibly to either the site 3 methine or the site 7 methylene. While we might be tempted to assume that H5 will couple to both H6's, we should bear in mind that the geometry of taxadiene in the region where the C3-C8 six-membered ring with its one double bond is such that H5 may not couple with the pseudoaxial ^{1}H on C6. That is, the nearly 90 degree dihedral angle between H5 and H6$_{ax}$ will produce little, if any, J-coupling. The cross peak between H5 and the ^{1}H signal 1.95 ppm is attributed to coupling to the equatorial H6 (pro-S). We write for site 6: $\delta_H = 1.95$ & 2.12 ppm and $\delta_C = 24.0$ ppm.

The remaining H5-participating COSY cross peak can either involve the H3 methine or the H7 methylene group. Based on the COSY data alone, the definitive assignment is not possible.

TABLE 4.1.3 ^{1}H and ^{13}C NMR Signals of Taxadiene Listed by Group Type

^{1}H Signal (ppm)	^{13}C Signal (ppm)	Group Type
–	138.7	Nonprotonated
–	138.2	Nonprotonated
–	130.0	Nonprotonated
5.39	122.0	Methine
1.68	45.0	Methine
1.41 & 1.82	42.1	Methylene
2.63	40.5	Methine
–	39.6	Nonprotonated
1.20 & 1.76	39.2	Methylene
–	38.0	Nonprotonated
1.10	31.4	Methyl
1.80 & 2.26	30.4	Methylene
1.62 & 1.72	29.1	Methylene
1.33	26.9	Methyl
2.16 & 2.63	25.2	Methylene
1.73	24.7	Methyl
1.95 & 2.12	24.0	Methylene
1.59 & 2.04	23.4	Methylene
0.93	22.3	Methyl
1.62	21.9	Methyl

If we examine the gHMBC spectrum, we observe that the H5 resonance at 5.39 ppm shares a strong correlation with a methine ^{13}C signal at 40.5 ppm which, according to the seventh row of Table 4.1.3, is the ^{13}C signal of a methine group whose ^{1}H signal is observed at 2.63 ppm. Therefore we can now assert that the H5 correlation to the signal at 2.63 ppm must be to the site 3 methine ^{1}H. We write for site 3: $\delta_H = 2.63$ ppm and $\delta_C = 40.5$ ppm.

Knowing the chemical shift of the site 20 methyl ^{1}H's, we can readily observe in the gHMBC spectrum that the narrow resonance from the H20's at 1.73 ppm correlates only with the ^{13}C signal at 138.7 ppm, which is only one of the three nonprotonated sp^2-hybridized carbon signals. This nonprotonated ^{13}C signal is assigned to site 4: $\delta_C = 138.7$ ppm.

Having now assigned two of the three methine signals (H3 and H5), the remaining methine signal must be that of site 1. We write for site 1: $\delta_H = 1.68$ ppm and $\delta_C = 45.0$ ppm.

The ^{1}H NMR signals from the geminal methyl groups at sites 16 and 17 share a moderate ^4J-spawned cross peak in the COSY spectrum. No other pair of methyl chemical shifts share a cross peak as pronounced as that observed between the ^{1}H signals at 1.10 and 1.33 ppm.

At this point we cannot determine which signal arises from which site, but the signals from the geminal methyl groups have been identified. In the gHMBC spectrum, the geminal methyl ^{1}H signals at 1.10 and 1.33 ppm are expected to correlate with the ^{13}C signals from C1, C11, and C15. As expected, the geminal methyl ^{1}H signals correlate in the gHMBC spectrum with the C1 signal at 45.0 ppm, confirming our site 1 assignment. The H16 and H17 signals also share gHMBC cross peaks with the signals from nonprotonated ^{13}C's at 138.2 and 39.6 ppm. The ^{13}C signal at 138.2 ppm must arise from the sp^2-hybridized C11. We write for site 11: $\delta_C = 138.2$ ppm. We assign the 39.6 ppm ^{13}C signal to the attachment point of the geminal methyl groups, site 15: $\delta_C = 39.6$ ppm.

Having now identified the NMR signals from three of the five methyl groups (16, 17, 20), we can differentiate between the remaining two methyl groups (18 and 19) by noting that the methyl at site 19 is bound to an sp^3-hybridized carbon at site 8 while the methyl at site 18 is bound to an sp^2-hybridized carbon at site 12. As we might expect, the more downfield of the two remaining methyl ^{1}H signals (at 1.62 ppm) correlates with two downfield ^{13}C signals (C11 and C12), while the more upfield of the two remaining methyl ^{1}H signals (at 0.93 ppm) only correlates with upfield ^{13}C signals. Therefore the methyl ^{1}H signal at 1.62 ppm is assigned to site 18 and the methyl ^{1}H signal at 0.93 ppm is assigned to site 19. We write for site 18: $\delta_H = 1.62$ ppm and $\delta_C = 21.9$ ppm and for site 19: $\delta_H = 0.93$ ppm and $\delta_C = 22.3$ ppm.

In the gHMBC spectrum, the H18 signal at 1.62 ppm shares a correlation with the C11 signal at 138.2 ppm and also with the signal from a nonprotonated ^{13}C at 130.0 ppm, which must be from C12. We write for site 12: $\delta_C = 130.0$ ppm.

The H19 signal at 0.93 ppm participates in four strong gHMBC cross peaks in the upfield ^{13}C chemical shift region. These cross peaks are attributed to coupling to C8, which is two bonds from the H19's, and to C3, C7, and C9 which are all three bonds distant from the H19's. The ^{13}C chemical shifts for these four gHMBC cross peaks are observed at 42.1, 40.5 (already know to be C3), 39.2, and 38.0 ppm. Of these four ^{13}C resonances, only the one at 38.0 ppm is known to correspond to that of a nonprotonated carbon site (see Table 4.1.3), and so this ^{13}C signal is assigned to site 8: $\delta_C = 38.0$ ppm. We therefore also now know that the methylenes at sites 7 and 9 have ^{13}C shifts of 42.1 and 39.2 ppm, but we do not know which is which.

Having now assigned all methyl, methine, and nonprotonated sites, we have only to finish the assignment of the methylene sites.

The well-resolved ^{1}H signal at 1.20 ppm, known to be from a methylene group, is observed to generate cross peaks with the downfield C5 signal at 122.0 ppm in the gHMBC spectrum, as well as with the C3 signal at 40.5 ppm, with the C8 signal at 38.0 ppm, with the C6 signal at 24.0 ppm, and with the C19 signal at 22.3 ppm. This ^{1}H signal at 1.20 ppm is therefore attributed to the methylene group at site 7: $\delta_H = 1.20$ & 1.76 ppm and $\delta_C = 39.2$ ppm. Sites 2 and 9 are easy to rule out. The H2 and H9 signals will share gHMBC cross peaks with a different set of the ^{13}C resonances whose shifts we already know. The H7 signal at 1.20 ppm also shares a strong gHMBC cross peak with the methylene ^{13}C signal at 42.1 ppm. The closest methylene group is found at site 9. C9 is three bonds distant from the H7's, and so we assign the ^{13}C signal at 42.1 ppm to site 9: $\delta_H = 1.41$ & 1.82 ppm and $\delta_C = 42.1$ ppm.

Gratifyingly, one of the tentatively assigned H9 signals is well isolated along the ^{1}H chemical shift axis at 1.41 ppm. This ^{1}H resonance shares gHMBC correlations with the C11 signal at 138.2 ppm, with the C3 signal at 40.5 ppm, with the C7 signal at 39.2 ppm, with the C8 signal at 38.0 ppm, with an unidentified methylene ^{13}C signal at 25.2 ppm, and with the C19 signal at 22.3 ppm. The unassigned methylene ^{13}C signal at 25.2 ppm is assigned to site 10, because the signal of every other carbon two or three bonds distant from H9 shares a gHMBC cross peak with the H9 resonance. We write for site 10: δ_H = 2.16 & 2.63 ppm and δ_C = 25.2 ppm. The site 10 assignment is consistent with gHMBC cross peaks shared between the two H10 signals at 2.16 & 2.63 ppm and the C11 signal at 138.2 ppm, the C9 signal at 42.1 ppm, and the C8 signal at 38.0 ppm. Note the presence of an interfering cross peak from the $^2J_{CH}$ between H3 whose signal is observed at 2.63 ppm and C8 whose signal is observed at 38.0 ppm. Interestingly, only the pro-S (up) ^{1}H at site 10 generates a signal that shares a gHMBC cross peak with the C15 signal at 39.6 ppm.

There are three unassigned methylene groups, corresponding to ^{1}H signals at 1.80 & 2.26 ppm, at 1.62 & 1.72 ppm, and at 1.59 & 2.04 ppm. Only the second set of methylene ^{1}H's, with chemical shifts of 1.62 & 1.72 ppm, are observed to both share gHMBC cross peaks with the C3 signal at 40.5 ppm. These methylene ^{1}H signals are assigned to site 2: δ_H = 1.62 & 1.72 ppm and δ_C = 29.1 ppm. Although the H1 signal at 1.68 ppm is observed to generate a confirming gHMBC cross peak with the C2 signal at 29.1 ppm, the H3 resonance fails to share a similar gHMBC cross peak with the C2 signal.

The last two unassigned methylene groups are found at sites 13 and 14. The methylene ^{1}H signals at 1.80 & 2.26 ppm are observed to share gHMBC cross peaks with the C11 signal at 138.2 ppm. These methylene ^{1}H signals are assigned to site 13: δ_H = 1.80 & 2.26 ppm and δ_C = 30.4 ppm.

By the process of elimination, the last set of methylene NMR signals is assigned to site 14: δ_H = 1.59 & 2.04 ppm and δ_C = 23.4 ppm. We are gratified to note the presence of confirming gHMBC cross peaks between the H14 signals at 1.59 & 2.04 ppm and the C2 signal at 29.1 ppm. It is also noted that the C15 signal fails to correlate with both the ^{1}H signals of either sites 13 or 14, and so the C15 signal is of no utility in confirming our assignment of sites 13 versus 14.

The final assignment we must make is to determine which methyl signals should be attributed to sites 16 and 17. We examine the gHMBC spectrum to find evidence of a ^4J from H2R to C16. The gHMBC spectrum is disappointing in that the cross peaks from the H2 signals at 1.62 & 1.72 ppm appear to both correlate equally well in the gHMBC spectrum to the geminal methyl ^{13}C signals at 26.9 and 31.4 ppm. The second method we have available to determine which methyl NMR signal pair is attributable to site 16 versus site 17 is the 2-D ^{1}H-^{1}H NOESY NMR spectrum (Figs. 4.1.11 and 4.1.12). The methyl ^{1}H signal at 1.10 ppm is observed to share NOESY correlations with the H13 signal at 2.26 ppm and with the H14 signal at 2.04 ppm. We assign the 1.10 ppm methyl signal to site 16: δ_H = 1.10 ppm and δ_C = 31.4 ppm. The methyl ^{1}H signal at 1.33 ppm happily correlates with spins on the opposite face of the plane defined by the bonds that C15 shares with C1 and C11. It should be noted that the rendering of the taxadiene structure in Fig. 4.1.1 is somewhat misleading: The site 16 methyl group should be viewed as coming up out of the plane of the figure and pointing toward site 13. Fig. 4.1.13 contains a 2-D rendering of the 3-D structure of taxadiene with the chemical shifts near each atom allowing the reader a better understanding of the observed NOE contacts. The methyl ^{1}H signal at 1.33 ppm shares NOESY cross peaks with the H10 signal at 2.16 ppm, with the H9 signal at 1.82 ppm, and with the H2 signal at 1.72 ppm. The 1.33 ppm methyl ^{1}H signal is therefore assigned to site 17: δ_H = 1.33 ppm and δ_C = 26.9 ppm.

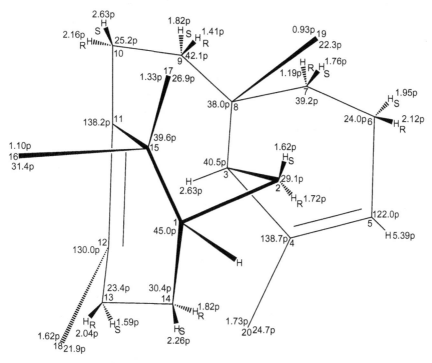

FIG. 4.1.13 The structure of taxadiene with assigned chemical shifts appearing at the molecular sites, drawn to show the short interatomic distances that generate observed NOESY cross peaks.

4.2 ISOOXACYCLOTAXANE IN BENZENE-D_6

Steven Edgar provided the author with a sample of isooxacyclotaxane (isoOCT) dissolved in benzene-d_6 (benzene-d_6, more often than not, provides better chemical shift dispersion than does chloroform-d). The empirical formula of the isoOCT was known to be $C_{20}H_{32}O$.

Using the integrals of the 1-D ^{1}H NMR spectrum, it was of course possible to determine how the NMR signals of the 32 ^{1}H's in the isoOCT molecule are distributed along the ^{1}H chemical shift axis. By using the phase-sensitive 2-D ^{1}H-^{13}C HSQC spectrum, the signals from each methyl, methylene, and methine group were identified. For overlapped regions of the 1-D ^{1}H NMR spectrum, chemical shifts were measured directly from the 2-D ^{1}H-^{13}C HSQC NMR spectrum, but this methodology fails to employ the "center-of-gravity" measurement (point at which 50% of the integrated signal intensity is found) of a ^{1}H resonance and instead lists the mid-point of the multiplet as its chemical shift position. The chemical shifts of ^{13}C's were determined via a 1-D ^{13}C NMR spectrum and were paired up to HSQC cross peaks.

As a general rule, measurement of ^{13}C chemical shift via the HSQC spectrum is far less reliable than measurement using the 1-D ^{13}C NMR spectrum, and still somewhat less reliable than using the gHMBC spectrum. One might be tempted to argue that the higher sensitivity of the HSQC experiment compared with that of the gHMBC experiment makes it more appropriate to use the HSQC spectrum for the measurement of ^{13}C peak positions, but recall that HSQC signal detection involves the decoupling of twenty to thirty kilohertz of the ^{13}C chemical shift axis which in turn often heats up the sample. Sample heating causes various resonances to deviate significantly from the position in which they are observed using NMR methods not prone to RF heating, i.e., the 1-D ^{13}C spectrum with ^{1}H decoupling occurring over a much more modest range of perhaps five kilohertz. Therefore in cases where there is too little material to observe the 1-D ^{13}C NMR signal, our best course of action is to measure ^{13}C shifts directly from the cross peaks in the gHMBC and not from those in the HSQC spectrum. Additionally, nonprotonated carbons fail to generate cross peaks in the HSQC spectrum.

The 1-D ^{1}H NMR spectrum of isoOCT appears in Fig. 4.2.1. The 1-D ^{13}C NMR spectrum of isoOCT is found in Fig. 4.2.2. The 2-D ^{1}H-^{1}H COSY NMR spectrum is shown in Fig. 4.2.3. The 2-D ^{1}H-^{13}C HSQC NMR spectrum of isoOCT is displayed in Fig. 4.2.4, and the 2-D ^{1}H-^{13}C gHMBC NMR spectrum in Fig. 4.2.5.

Examination of the 1-D ^{13}C NMR spectrum for signals not found to participate in HSQC cross peaks was used to identify five nonprotonated carbon signals. The gHMBC spectrum was able to confirm that the spins responsible for these signals were involved in long-range coupling to [1]H's elsewhere in the molecule, thus allowing us to identify any potential artifacts and/or impurities to distinguish them from the signals from nonprotonated [13]C's in the isoOCT molecule.

Table 4.2.1 was generated based on the observation of the signals in the 1-D ^{1}H, 2-D ^{1}H-^{13}C HSQC, and 1-D ^{13}C NMR spectra. A lower case letter is assigned to the NMR signals of each carbon site in isoOCT. The chemical shift of the NMR signal from each of the 20 ^{13}C environments of isoOCT is paired with the chemical shift of the NMR signal from directly attached [1]H's, if extant.

It should be noted that all the hydrogens present in the isoOCT molecule were directly bonded to ^{13}C atoms, but this is not always the case, so it is important to account for every ^{1}H signal, including those that may arise from [1]H's bound to heteroatoms such as nitrogen and oxygen. In this case, the sole oxygen atom is observed to withdraw electron density from two carbons. To wit, we observe a nonprotonated ^{13}C signal at 97.6 ppm (label p) and a methine ^{13}C signal at 77.1 ppm (label m). Thus we are confident that the oxygen atom is present as part of an ether (C-O-C) linkage.

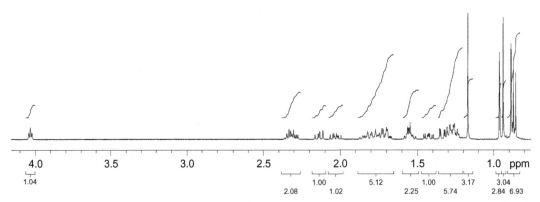

FIG. 4.2.1 The 1-D ^{1}H NMR spectrum of isoOCT in benzene-d_6. *Reprinted with permission from Edgar, S., Zhou, K., Qiao, K., King, J.R., Simpson, J.H., Stephanopoulos, G., 2016. Mechanistic insights into taxadiene epoxidation by taxadiene-5α-hydroxylase. ACS Chem. Biol. 11 (2), 460–469. Copyright 2015, American Chemical Society.*

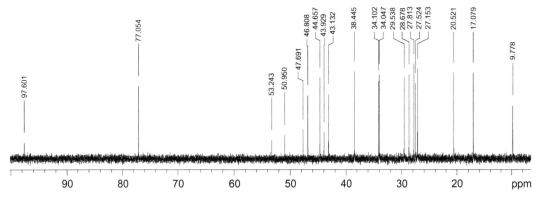

FIG. 4.2.2 The 1-D ^{13}C NMR spectrum of isoOCT in benzene-d_6. *Reprinted with permission from Edgar, S., Zhou, K., Qiao, K., King, J.R., Simpson, J.H., Stephanopoulos, G., 2016. Mechanistic insights into taxadiene epoxidation by taxadiene-5α-hydroxylase. ACS Chem. Biol. 11 (2), 460–469. Copyright 2015, American Chemical Society.*

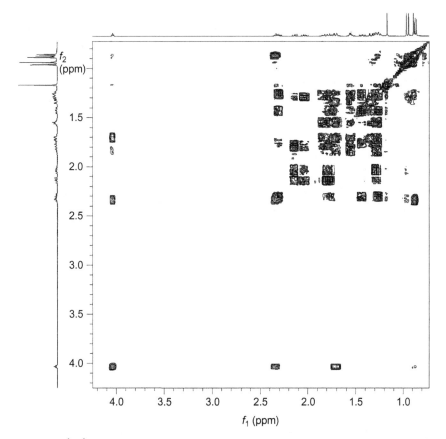

FIG. 4.2.3 The 2-D ^{1}H-^{1}H COSY NMR spectrum of isoOCT in benzene-d_6. *Reprinted with permission from Edgar, S., Zhou, K., Qiao, K., King, J.R., Simpson, J.H., Stephanopoulos, G., 2016. Mechanistic insights into taxadiene epoxidation by taxadiene-5α-hydroxylase. ACS Chem. Biol. 11 (2), 460–469. Copyright 2015, American Chemical Society.*

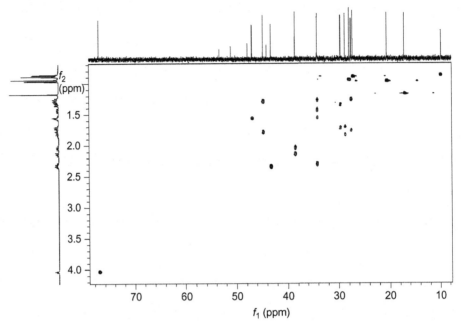

FIG. 4.2.4 The 2-D ^{1}H-^{13}C HSQC NMR spectrum of isoOCT in benzene-d_6. *Reprinted with permission from Edgar, S., Zhou, K., Qiao, K., King, J.R., Simpson, J.H., Stephanopoulos, G., 2016. Mechanistic insights into taxadiene epoxidation by taxadiene-5α-hydroxylase. ACS Chem. Biol. 11 (2), 460–469. Copyright 2015, American Chemical Society.*

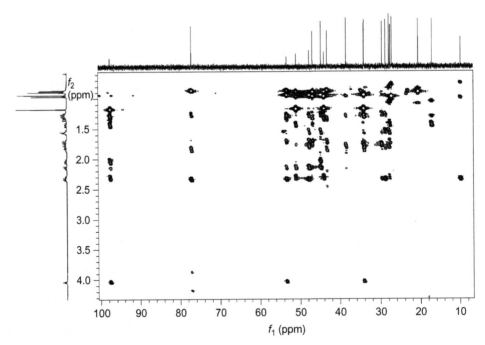

FIG. 4.2.5 The 2-D ^{1}H-^{13}C gHMBC NMR spectrum of isoOCT in benzene-d_6. *Reprinted with permission from Edgar, S., Zhou, K., Qiao, K., King, J.R., Simpson, J.H., Stephanopoulos, G., 2016. Mechanistic insights into taxadiene epoxidation by taxadiene-5α-hydroxylase. ACS Chem. Biol. 11 (2), 460–469. Copyright 2015, American Chemical Society.*

TABLE 4.2.1 Observed ^1H and ^{13}C Shifts of the Methyl, Methylene, Methine, and Nonprotonated Carbons of isoOCT

Total Number Present	Carbon Type	Letter Labels	Shifts (^1H <& ^1H>, ^{13}C) or (^{13}C) (in ppm)
5	Methyl (CH$_3$)	a, b, c, d, e	a (0.95, 27.8), b (0.89, 27.2), c (0.97, 20.5), d (1.17, 17.1), e (0.87, 9.8)
7	Methylene (CH$_2$)	f, g, h, i, j, k, l	f (1.29 & 1.78, 44.7), g (2.03 & 2.13, 38.4) h (1.43 & 2.30, 34.1), i (1.26 & 1.55, 34.0) j (1.35 & 1.72, 29.5), k (1.71 & 1.83, 28.7) l (1.255 & 1.76, 27.5)
3	Methine (CH)	m, n, o	m (4.03, 77.1), n (1.56, 46.8), o (2.34, 43.1)
5	Nonprotonated (C)	p, q, r, s, t	p (97.6), q (53.2), r (51.0), s (47.7), t (43.9)

Based on the fact that all of the observed ^1H signal and all save two of the ^{13}C signals have chemical shifts that are in the aliphatic chemical shift range, we proceed assuming that we have a lone ether linkage in the molecule, no carbon-carbon double bonds, and therefore only rings to satisfy the octet rule for all of our carbon atoms. For isoOCT with its empirical formula of $C_{20}H_{32}O$, the index of hydrogen deficiency (IHD) is calculated as follows:

$$\text{IHD of } C_{20}H_{32}O : \left(2 \times \# : \text{carbons} + 2 - \# : \text{hydrogens}\right) / 2 = \left(20 \times 2 + 2 - 32\right) / 2 = 5$$

Taxadiene itself also has an IHD of five, but two double bonds mean that only three cyclic structures are present. For isoOCT, *five* rings must be found in the correct structure. The number of possible 5-ringed structures one can generate with a $C_{20}O$ skeleton is alarmingly large.

Initial attempts at deducing the structure of isoOCT were carried out by the author using pencil and paper. The three-dimensionality of the building blocks for the molecule—i.e., sp^3-hybridized carbon atoms—were found to be incompatible with the 2-D renderings on paper which the author utilized in his initial attempts to elucidate the structure of isoOCT. In this realm of fuzzy logic there is no single correct approach, insofar as we judge the validity of the path taken based on whether or not it ultimately allows us to arrive at the correct destination. The method used most effectively by the author was to obtain a molecular model kit and label each methyl, methylene, methine, and nonprotonated carbon atom as an isolated fragment. This was done by taping labels with carbon type and the ^{13}C and ^1H chemical shift(s) to each of 20 carbons from the model kit. Once prepared with all of the known fragments, it was possible to then piece the fragments together with one another in an attempt to satisfy all of the coupling relationships observed in the 1-D ^1H spectrum the 2-D ^1H-^1H COSY spectrum, and the 2-D ^1H-^{13}C gHMBC spectrum.

Although one might be tempted to attempt to translate observed spectral signals (1-D ^1H and ^{13}C signals, as well as 2-D cross peaks) directly into bonded relationships between fragments of the model, it is important to use a sheet of paper as an intermediary between NMR signals and model construction. Using a paper intermediary allows one to undo one or more steps if an unsatisfactory result appears imminent, whereas with just the model in one's hands, one wrong pull and model's utility is undone.

For the author, three attempts were required, on three separate days, to solve the structure of isoOCT, and even for the third and apparently correct structure, there was still an error that remained undetected until the details of the mechanism were established the following day.

Overlap of the ^{1}H signals from the methylene and methine groups along the ^{1}H chemical shift axis (see the 1-D ^{1}H NMR spectrum in Fig. 4.2.1) render many portions of the 2-D ^{1}H-^{1}H COSY spectrum in Fig. 4.2.3 largely unusable for the structure-determination portion of this project. Only in retrospect was it possible to correlate the myriad COSY cross peaks with their proper two- and three-bond couplings. Because seven methylene groups are present and three methine groups, homonuclear coupling broadens their ^{1}H signals into multiplets that overlap extensively, diminishing the utility of the COSY spectrum.

For overlapped ^{1}H signals, we use the COSY spectrum for confirmation of couplings predicted in hindsight, not for establishing unambiguous spin proximities in our initial phase which must involve only high confidence assignments. We first exhaust all of our high confidence assignments, leaving hopefully only a select few possibilities that we may have to address using trial and error.

On the other hand, the 2-D ^{1}H-^{13}C gHMBC spectrum enjoys greater signal dispersion and therefore contains what we might suppose would be much more useful cross peaks. Using the gHMBC spectrum, however, we cannot determine if a given cross peak is the result of a $^2J_{CH}$ versus a $^3J_{CH}$. The conformational rigidity placed upon the isoOCT molecule by its five cyclic connections complicates the problem of establishing geminal versus vicinal proximity of ^{1}H's and ^{13}C's by locking in a narrow range of molecular conformations that in turn prevents free rotation from averaging out $^3J_{CH}$'s. Put more simply, molecular rigidity results in observable $^3J_{HC}$ couplings that are easy to mistake for $^2J_{HC}$'s.

The structure determination is initiated at the ether linkage from the most downfield methine group, the label m (4.03, 77.1) methine, to the sole oxygen and then to the most downfield nonprotonated carbon with a ^{13}C chemical shift of 97.6 ppm (label p). The fragment is shown in Fig. 4.2.6.

The 2-D ^{1}H-^{1}H COSY spectrum contains a strong cross peak between ^{1}H resonances at 4.03 (label m) and 2.34 ppm (label o), so the label o (2.34, 43.1) methine group is attached to the label m (4.03, 77.1) methine group. Additionally, the ^{1}H signal of the label m methine at 4.03 ppm is observed to correlate in the COSY spectrum with both of the label k methylene ^{1}H signals at 1.71 & 1.84 ppm. Because the label m methine ^{1}H signal correlates with the signals of both ^{1}H's on the label k methylene group, the label k methylene is placed adjacent to the label m methine. If the label m methine ^{1}H signal only correlated with one of the two methylene ^{1}H signals, we would be more reluctant to place the methylene adjacent to the methine. Our progress thus far is shown in Fig. 4.2.7.

Next, we place the label e (0.87, 9.8) methyl group next to the label o (2.34, 43.1) methine group using the COSY cross peak observed between these two resonances. Note in the 1-D ^{1}H NMR spectrum that the label e methyl resonance is split into a doublet through coupling

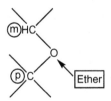

FIG. 4.2.6   The first logical step showing how NMR evidence couples isoOCT carbons labeled m and p across the ether linkage.

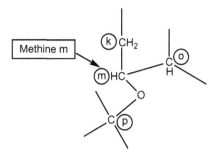

FIG. 4.2.7 The second logical step showing how NMR evidence couples isoOCT carbons labeled o and k to the existing framework.

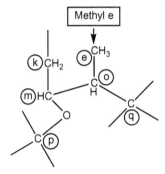

FIG. 4.2.8 The third logical step showing how NMR evidence couples isoOCT carbons labeled e and q to the existing framework.

to the label o methine ^{1}H. The label e methyl ^{1}H signal at 0.87 ppm shares confirming gHMBC correlations with the label m methine ^{13}C signal at 77.1 ppm and with the label o methine ^{13}C signal at 43.1 ppm. The label e methyl ^{1}H signal also correlates in the gHMBC spectrum with the label q nonprotonated ^{13}C signal at 53.2 ppm. We place the label q nonprotonated ^{13}C group on the label o methine. Fig. 4.2.8 depicts the point to which we have developed our molecular fragment.

It is noted that the remaining four methyl groups are all bound to nonprotonated carbons, as only the label e methyl signal at 0.87 ppm is observed to be split into a doublet, indicating the proximity of the methyl ^{1}H's to a methine group. The other four methyl group ^{1}H signals are all singlets and so the remaining four methyl groups all must reside on nonprotonated carbon sites.

We now leave the molecular fragment we have developed thus far and work to develop connections in a different portion of the isoOCT molecule. Taxadiene contains a pair of geminal methyl groups. Since isoOCT is derived from taxadiene, it is reasonable to suppose that isoOCT might also contain geminal methyl groups. The label b methyl ^{1}H signal at 0.89 ppm and the label c methyl ^{1}H signal at 0.97 ppm both correlate in the gHMBC spectrum with the label r nonprotonated ^{13}C signal at 51.0 ppm, with the label n methine ^{13}C signal at 46.8 ppm, and with the label t nonprotonated ^{13}C signal at 43.9 ppm. The similarity in coupling partners for the label b and c methyl groups indicates that these are the geminal methyl groups, and so we assert that labels b and c correspond to geminal methyl groups. The gHMBC correlations from the label b and c methyl ^{1}H signals to the label t ^{13}C signal are more intense than to the label r ^{13}C signal. We therefore attach the geminal methyl groups to the nonprotonated carbon generating the label t NMR signal. We can also employ a simple chemical shift argument to assert that two methyl groups are more likely to be bound to the ^{13}C with the more upfield (lower ppm) chemical shift, as the other nonprotonated ^{13}C signal must result from a carbon site with fewer than two electron-density-donating methyl groups attached. We place the label r nonprotonated carbon on one side of the label t nonprotonated carbon, and the label n methine group on the other side. This fragment is shown in Fig. 4.2.9.

Continuing to utilize methyl group signals, we observe that the label d methyl ^{1}H signal at 1.17 ppm correlates in the gHMBC with the label p nonprotonated ^{13}C signal at 97.6 ppm, with the label r nonprotonated ^{13}C signal at 51.0 ppm, with the label t nonprotonated ^{13}C signal at 43.9 ppm, and with the label h methylene ^{13}C signal at 34.1 ppm. Having already located the spins associated with labels p, t, and r, we are able to assert that the label r nonprotonated carbon is the attachment point of the label d methyl group, because its ^{13}C NMR signal is more upfield from the label p nonprotonated ^{13}C signal. This allows us to place the label h methylene group on the label r nonprotonated carbon. Putting these discoveries into a visual format, we arrive at Fig. 4.2.10.

FIG. 4.2.9   The fourth logical step showing how NMR evidence couples isoOCT carbons labeled b, c, n, r, and t together to form a new molecular fragment.

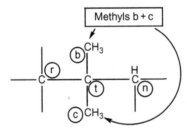

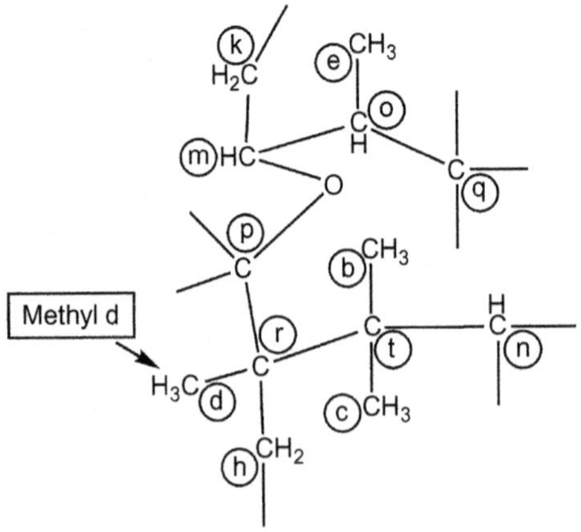

FIG. 4.2.10   The fifth logical step showing how NMR evidence couples the isoOCT molecular fragments of steps three and four together using the carbons labeled d and h.

The last methyl group generates the label a signals (0.95, 27.8). In the gHMBC spectrum, the label a methyl ^{1}H signal is observed to correlate with the already located label q nonprotonated ^{13}C at 53.2 ppm, with the label s nonprotonated ^{13}C at 47.7 ppm, with the label f methylene ^{13}C at 44.7 ppm, and with the label i methylene ^{13}C at 34.0 ppm. The label a methyl ^{1}H signal is a singlet, so the label a methyl group must be attached to one of the two nonprotonated ^{13}C's: label q or s. Since the label q nonprotonated carbon has already been placed alpha to the label o methine group, we place the label a methyl group on the label s nonprotonated carbon, as placing the label a methyl group on the label q nonprotonated carbon would likely generate a correlation between the label a methyl ^{1}H signal at 0.95 ppm and the label o methine ^{13}C signal at 43.1 ppm (not observed). The label f methylene ^{1}H's whose signals are found at 1.29 & 1.78 ppm are observed to couple in the COSYspectrum with the well-resolved label g methylene ^{1}H signals whose chemical shifts are 2.03 & 2.13 ppm. In the gHMBC spectrum, the label g methylene ^{1}H signals at 2.03 & 2.13 ppm correlate with the label f ^{13}C signal at 44.7 ppm, but also with the label p nonprotonated ^{13}C signal at 97.6 ppm. In hindsight we rationalize that the label g methylene ^{1}H signals are well resolved because they are anomalously far downfield (at higher chemical shift values) due to their proximity to the oxygen atom on the other side of the label p nonprotonated carbon. We have closed a ring involving the groups labeled p, g, f, s, q, o, m, and the oxygen atom. We will make four more of these ring closure steps. We codify the information gleaned from the COSY and gHMBC correlations of the label f and g methylenes as well as from the gHMBC correlations involving the label a methyl group ^{1}H signal into the visual format shown in Fig. 4.2.11.

At this point, we have only two remaining single carbon groups to attach to our expanding fragment: the label j and l methylene groups. Additionally, we must of course satisfy the

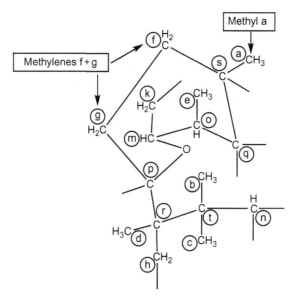

FIG. 4.2.11 The sixth logical step showing how NMR evidence couples isoOCT carbons labeled a, f, g, and s to the existing framework, thereby closing the first ring.

octet rule by finding bonding partners for the label i methylene, the label n methine (two), the label h methylene, the label k methylene, the label p nonprotonated carbon, and the label q nonprotonated carbon (two).

There are five overlapping ¹H resonances in the chemical shift range from 1.7 to 1.9 ppm, with the already located label k methylene ¹H signals located at the extrema of this 0.2 ppm range. The label k methylene ¹H signals at 1.71 & 1.83 ppm are observed to share COSY correlations with ¹H signals at 1.55/6 ppm and at 1.25/6 ppm. The label i methylene group generates ¹H signals with chemical shifts of 1.26 & 1.55 ppm, and so we connect the label i methylene group to the label k methylene group, thereby closing another ring. We observe confirming gHMBC cross peaks between the label i ¹H signals at 1.26 & 1.55 ppm and the label k ¹³C signal at 28.7 ppm, as well as between the label k ¹H signals at 1.71 & 1.83 ppm and the label i ¹³C signal at 34.0 ppm. Our progress thus far is shown in Fig. 4.2.12.

The label h methylene ¹H resonances at 1.43 & 2.30 ppm are observed to share COSY cross peaks with the label l methylene ¹H resonances at 1.255 & 1.76 ppm. We place the label l methylene next to the label h methylene. We also observe gHMBC correlations between the well-resolved label h ¹H signal at 1.43 ppm and the following: the label r nonprotonated ¹³C signal at 51.0 ppm, the label n methine ¹³C signal at 46.8 ppm, the label l methylene ¹³C signal at 27.5 ppm, and the label d methyl ¹³C signal at 17.1 ppm. Of the four preceding gHMBC cross peaks, the connection to label n is new and allows us to place the label n methine group on the other side of the label l methylene group from the label h methylene group. Our result appears in Fig. 4.2.13.

The last methylene group is the label j methylene. The label j methylene ¹H signals at 1.35 & 1.72 ppm are observed to both share gHMBC cross peaks with the label q nonprotonated ¹³C signal at 53.2 ppm, with the label n methine ¹³C signal at 46.8 ppm, and with the label o methine ¹³C signal at 43.1 ppm. We reason that ¹³C's two bonds removed from methylene ¹H's will couple with both ¹H's, while ¹³C's that are three bonds from methylene ¹H's can

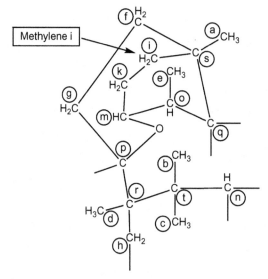

FIG. 4.2.12 The seventh logical step showing how NMR evidence couples the isoOCT carbon labeled i to the existing framework, thereby closing the second ring.

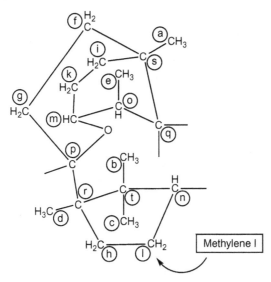

FIG. 4.2.13 The eighth logical step showing how NMR evidence couples the isoOCT carbon labeled l to the existing framework, thereby closing the third ring.

often have a zero $^3J_{CH}$ when a dihedral angle near 90 degrees is present. Therefore those ^{13}C signals that correlate with the signals of both ^1H's on a given methylene are most likely two bonds and not three bonds away. By only selecting those ^{13}C signals that share cross peaks with the signals of both ^1H's on a methylene group, we are able to (1) place high confidence in cross peak identities despite ^1H signal overlap and (2) locate ^{13}C's alpha to the methylene. Placement of the label j methylene group between the label n methine group and the label q nonprotonated carbon is consistent with the just-noted gHMBC cross peaks. We have closed another ring and now have only two loose ends to connect, namely, the label q nonprotonated carbon and the label p nonprotonated carbon. It comes as no surprise that we were not able to directly determine that two nonprotonated carbon sites were adjacent to one another, as we lack the INADEQUATE spectrum owing to its very low sensitivity. This also goes to show that we must mind our p's and q's. The progress we have made thus far is shown in Fig. 4.2.14.

Conversion of the 2-D molecule to the real molecule results in four distinct possibilities. Two of these possibilities are shown in Figs. 4.2.15 and 4.2.16. Mirror images of the molecules in these two figures account for the other two possible structures. The difference between the two shown structures involves mirroring the right half of the molecule, such that every group to the right of the plane roughly defined by the groups with labels p and q is horizontally mirrored.

The mechanism by which isoOCT is made from taxadiene is shown in Fig. 4.2.17. This mechanism was deduced by Stephen Edgar and reveals which of the two structures is the correct one. In step 1, taxadiene is oxidized to form an epoxide with the oxygen atom bridging the carbon-carbon double bond spanning sites 4 and 5 (Fig. 4.2.17A). Step 2 involves opening the epoxide ring to place the negatively charged oxygen on the site 5 carbon, and placing a balancing positive charge on site 4, making the first of four tertiary carbocation intermediates (Fig. 4.2.17B).

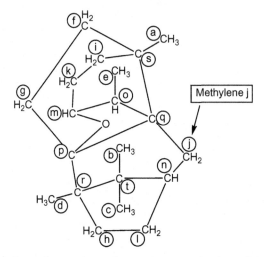

FIG. 4.2.14   The ninth logical step showing how NMR evidence couples the isoOCT carbon labeled j to the existing framework, thereby closing the fourth ring. Additionally, nonprotonated carbons with labels p and q are bonded to one another, closing the fifth ring.

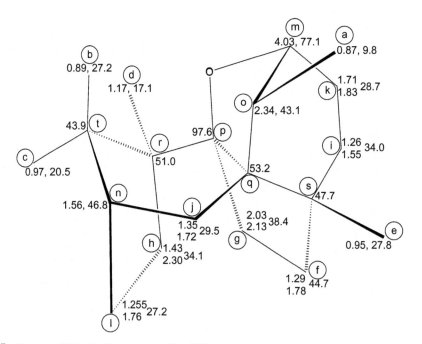

FIG. 4.2.15   One possibility for the structure of isoOCT.

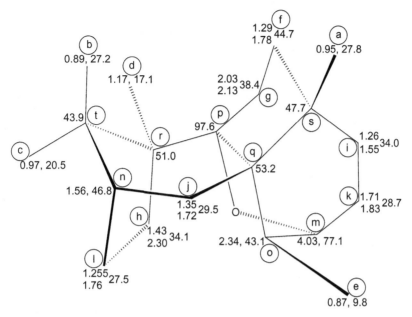

FIG. 4.2.16   A second possibility for the structure of isoOCT.

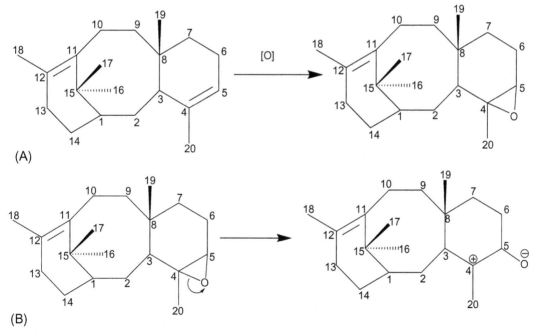

FIG. 4.2.17   The six mechanistic steps (A–F) that form isoOCT from its epoxide precursor.

(Continued)

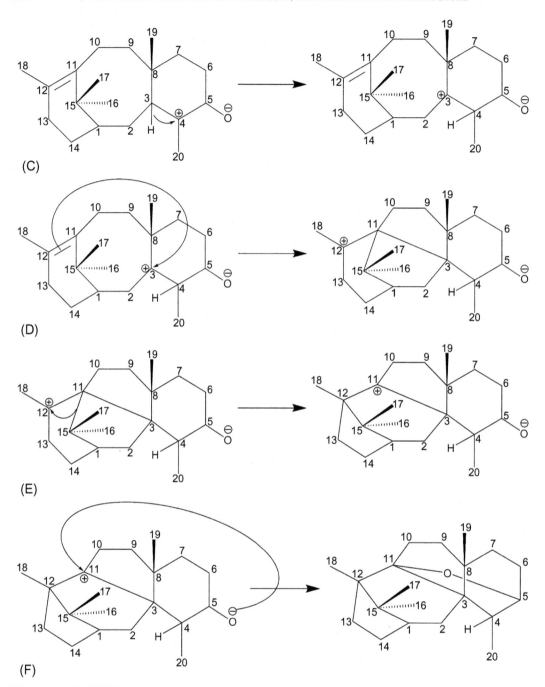

FIG. 4.2.17, CONT'D

In mechanistic step 3, a hydride migration occurs from the site 3 carbon to the site 4 carbon, placing the positive charge on the site 3 carbon (Fig. 4.2.17C). Step 4 is best understood while holding the model of the molecule, as the π electrons comprising the carbon-carbon double bond between sites 11 and 12 reach across the molecule to the site 3 carbocation center (Fig. 4.2.17D). The π electrons form a new bond connecting carbon 11 to carbon 3. The positive charge moves to the site 12 carbon.

In the next mechanistic step of the reaction, a tertiary carbocation rearrangement occurs. Mechanistic step 5 involves movement of the electron pair of the σ bond between carbons 11 and 15 to a position between carbons 12 and 15 (Fig. 4.2.17E). By shifting the carbocation from site 12 to site 11, it is possible that additional steric relief is afforded.

The last step of the reaction, step 6, the negatively charged oxygen atom on carbon 5 forms an ether linkage between carbon 5 and carbon 11 (Fig. 4.2.17F).

If we work through the preceding reaction mechanism with a model kit, we discover that only one of the two possible structures shown, that in Fig. 4.2.16, is consistent with the mechanism and the initial structure of taxadiene. We should always take the time to explore every avenue of discovery when we are attempting to elucidate the structure of a complex unknown molecule.

Bigger and More Complicated Molecules

5.1 A LARGE SYNTHETIC MOLECULE FOR LIQUID CRYSTALLINE APPLICATIONS (MOLECULE 5.1) IN CHLOROFORM-*D*

Drs. Jason Cox and Timothy Swager provided the author with a sample of the molecule shown in Fig. 5.1.1. The molecule has an empirical formula of $C_{38}H_{28}O_4$ and contains four aromatic rings, two pairs of triply bonded carbon atoms, and four sp^3-hybridized carbons that connect two of the aromatic rings to form one three-membered and two five-membered rings. The four oxygens are found in two ester linkages. This molecule generates six upfield (aliphatic) and 10 downfield (aromatic) ^1H resonances. Overlap in the downfield region of the ^1H chemical shift axis presents a formidable challenge as we progress through the assignment. The index of hydrogen deficiency for this molecule is a whopping 25, giving the molecule 18 nonprotonated carbon sites. Assigning the ^{13}C signals proves a challenge and, without the information contained in gHMBC spectrum, would not be possible.

All spectra were collected from a single sample provided to the author by Dr. Jason Cox (and his advisor, Dr. Timothy Swager) (Cox et al., 2013). The solvent was chloroform-*d*. It is noteworthy to mention that a real world sample such as this is a sample that will, for a spectroscopist like the author who does not synthesize samples himself, only be run once. The ^{13}C chemical shift window (f_1 dimension) in the gHMBC was set improperly, giving the result that the ester carbonyl signals appear in an aliased or folded position, wrapping around from the high ppm side of the gHMBC to low ppm values. These downfield ^{13}C signals are, with properly set spectral window parameters, observed at 168–169 ppm, but cross peaks involving these resonances appear at 14–15 ppm in the gHMBC spectrum due to the aforementioned mis-setting of the f_1 window.

The 1-D ^1H NMR spectrum of the molecule is shown in Fig. 5.1.2, and the 1-D ^{13}C NMR spectrum is shown in Fig. 5.1.3. An expanded portion of the 1-D ^{13}C NMR spectrum is also provided in Fig. 5.1.4 to allow a clearer assessment of the densely packed region of the ^{13}C chemical shift axis in the range from 115 to 141 ppm. The 2-D ^1H-^1H COSY NMR spectrum of molecule 5.1 appears in Fig. 5.1.5 and a second, expanded portion of the COSY spectrum

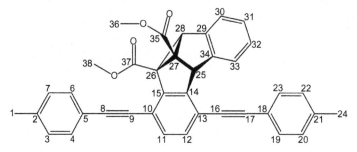

FIG. 5.1.1 The structure of molecule 5.1.

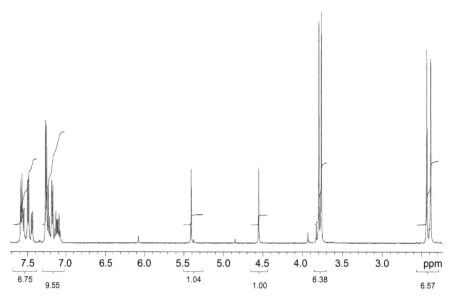

FIG. 5.1.2 The 1-D ^{1}H NMR spectrum of molecule 5.1 in chloroform-*d*.

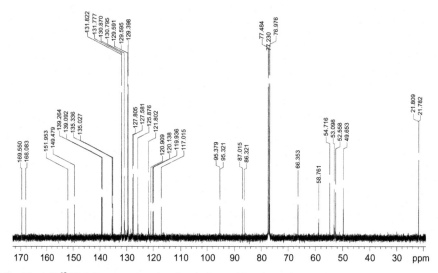

FIG. 5.1.3 The 1-D ^{13}C NMR spectrum of molecule 5.1 in chloroform-*d*.

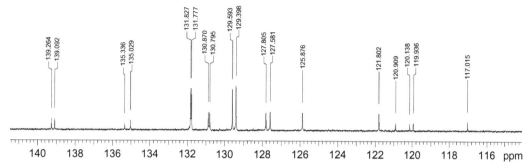

FIG. 5.1.4   An expanded portion (115–141 ppm) of the 1-D ^{13}C NMR spectrum of molecule 5.1 in chloroform-*d*.

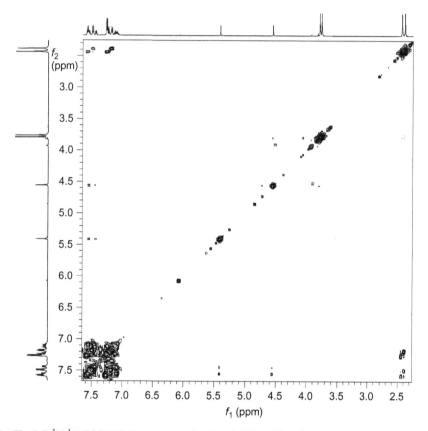

FIG. 5.1.5   The 2-D ^{1}H-^{1}H COSY NMR spectrum of molecule 5.1 in chloroform-*d*.

appears in Fig. 5.1.6. The 2-D ^{1}H-^{13}C HSQC NMR spectrum is found in Fig. 5.1.7, and the 2-D ^{1}H-^{13}C gHMBC NMR spectrum is shown in Fig. 5.1.8. An expanded portion of the gHMBC spectrum showing the most crowded downfield region appears in Fig. 5.1.9. As mentioned previously, the ^{13}C chemical shift range (f_1) in the gHMBC spectrum was set improperly such that correlations between ^{1}H's and the carbonyl ^{13}C's are folded and appear at 14–15 ppm.

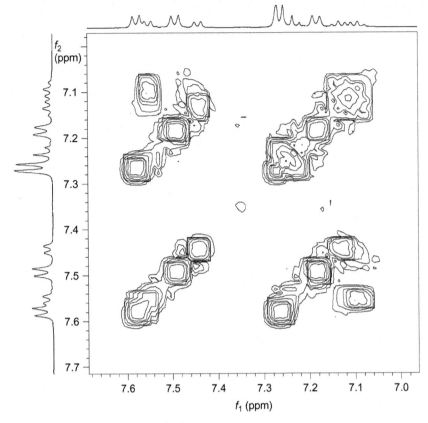

FIG. 5.1.6 An expanded portion of the 2-D ^{1}H-^{1}H COSY NMR spectrum of molecule 5.1 in chloroform-d.

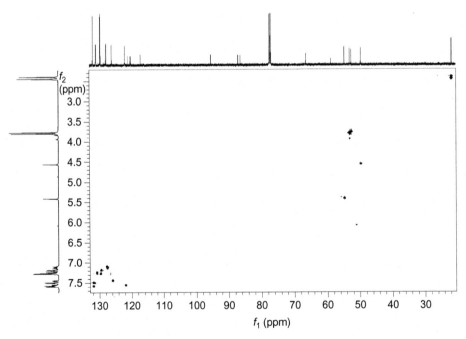

FIG. 5.1.7 The 2-D ^{1}H-^{13}C HSQC NMR spectrum of molecule 5.1 in chloroform-d.

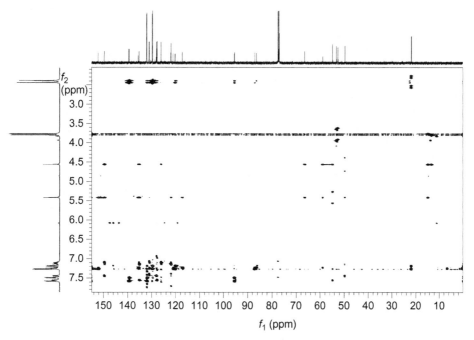

FIG. 5.1.8 The 2-D ^{1}H-^{13}C gHMBC NMR spectrum of molecule 5.1 in chloroform-*d*.

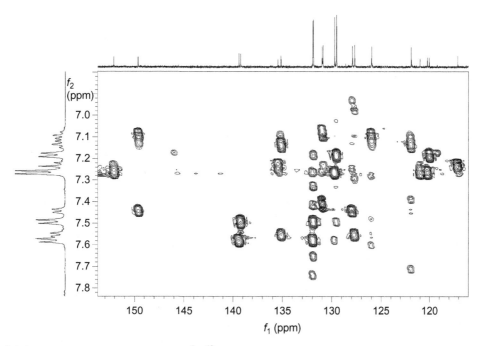

FIG. 5.1.9 An expanded portion of the 2-D ^{1}H-^{13}C gHMBC NMR spectrum of molecule 5.1 in chloroform-*d*.

The folded gHMBC correlations are fortuitously well resolved, not overlapping with any of the eight aliphatic ^{13}C signals.

Examination of the molecule's structure as it is shown in Fig. 5.1.1 allows us to predict multiplicities for the ^{1}H signal multiplets we anticipate observing, allowing us to correlate multiplet appearance with specific molecular sites containing hydrogen. Our predicted ^{1}H signal multiplicities appear in Table 5.1.1. Two sets of triplets (or doublets of doublets, if you are a multiplet prediction purist) are predicted for the signals of the ^{1}H's on sites 31 and 32 of the molecule—all the other aromatic ^{1}H resonances are expected to appear as doublets. We will surely make use of this unique attribute of two of the ^{1}H sites in the molecule as we proceed.

Table 5.1.2 contains an accounting of the various carbon types found in the molecule. An interesting feature of this molecule is that it has no methylene groups. We discern two c_2 rotational symmetry axes about the multiple bonds between C5 and C10 and also between C13 and C18 (they are collinear). These symmetry elements allow for rapid rotation of the two monofunctional aromatic rings, generating four pairs of doubly intense NMR signals in both the 1-D ^{1}H and 1-D ^{13}C NMR spectra.

Identification of the double intensity signals from sites 3/7, 4/6, 19/23, and 20/22 is complicated by spurious peak picking information in the 1-D ^{13}C NMR spectrum. Examination of the 1-D ^{13}C NMR spectrum allows us to readily locate the doubly intense aromatic ^{13}C signals.

TABLE 5.1.1   Predicted Multiplicities for the ^{1}H's of Molecule 5.1

Site in Molecule	Expected Multiplicity
1	s
3/7	d[a]
4/6	d[a]
11	d
12	d
19/23	d[a]
20/22	d[a]
24	s
25	s
28	s
30	d
31	t
32	t
33	d
36	s
38	s

[a]Double intensity.

TABLE 5.1.2 Carbons of Molecule 5.1

Type of Site	Site Number
CH_3 (methyl)	1, 24, 36, 38
CH_2 (methylene)	n/a
CH (methine)	3/7, 4/6, 11, 12, 19/23, 20/22, 25, 28, 30, 31, 32, 33
C_{np} (nonprotonated)	2, 5, 8, 9, 10, 13, 14, 15, 16, 17, 18, 21, 26, 27, 29, 34, 35, 37

We observe signal from a total of 14 aromatic ^{1}H's. The ^{1}H signals are distributed such that six are found in the chemical shift range from 7.4 to 7.7 ppm and eight in the range from 7.0 to 7.3 ppm. Because four of the signals correspond to symmetry doubled sites, we will observe 10 resonances and not 14: six normal- and four double-intensity.

In the more upfield ^{1}H chemical shift range of 7.0 to 7.3 ppm, we observe signal from eight ^{1}H's. The HSQC cross peaks involving the 7.0 to 7.3 ppm ^{1}H resonances correspond to two doubly represented and four singly represented ^{13}C sites.

In the more downfield ^{1}H chemical shift range of 7.4 to 7.7 ppm, we observe four HSQC cross peaks associated with what the integrals suggest are six ^{1}H signals. Two of the cross peaks must arise from the NMR signals of the symmetry-related sites (double intensity) and the other two from nonsymmetric (normal intensity) sites. The 1-D ^{1}H spectrum provides sufficient signal dispersion to allow us to determine that, in going from downfield to upfield (left to right), the intensities of the four doublet ^{1}H signals starting at 7.7 ppm are double (7.57 ppm), single (7.55 ppm), double (7.48 ppm), and single (7.44 ppm). We could try to identify the other two double intensity aromatic ^{1}H signals in the chemical shift range 7.0 to 7.3 ppm, but the COSY spectrum will provide us with an easier way to make this determination.

The six protonated ^{13}C signals corresponding to the four methyl groups and the two sp^3-hybridized methine groups are easy to locate using the HSQC spectrum. The two nonprotonated sp^3-hybridized ^{13}C signals from sites 26 and 27 are identified by the process of elimination. They do not participate in HSQC cross peaks and are found near 60 ppm, being found too far upfield to be considered as potential candidates for assignment as triple-bonded or aromatic ^{13}C signals.

Table 5.1.3 lists the observed ^{1}H and ^{13}C NMR signals from the molecule.

Although it is tempting to begin our assignments with methyl groups at the ends of the aromatic rings or the ester groups, we will instead focus on the two aliphatic methine sites 25 and 28. Otherwise we have an uncertainty we must propagate through many assignment steps before we discover which set of signals corresponds to which side of the molecule. Near the bottom of Table 5.1.3 (this is a big table spanning two pages) we find the rows listing the shifts of the signals from the two aliphatic methine ^{1}H/^{13}C pairs. The aliphatic methine ^{1}H resonance at 4.55 ppm is observed to share gHMBC cross peaks with both carbonyl ^{13}C signals at 169.6 and 168.1 ppm (folded to 15 and 14 ppm, respectively, on the f_1 axis of the gHMBC spectrum), while the other aliphatic methine ^{1}H signal at 5.42 ppm only correlates with the more downfield (169.8 aka 15 ppm) of the two carbonyl ^{13}C signals. The site 28 methine group is alpha to both carbonyl groups (sites 35 and 37) while the site 25 methine is only alpha to the site 35 carbonyl and beta to site 37. Because the four bonds between H25 and C37 are not

TABLE 5.1.3 ^{1}H and ^{13}C NMR Signals of Molecule 5.1 Listed by Group Type

^{1}H Signal (ppm)	^{13}C Signal (ppm)	Group Type
–	169.6	Nonprotonated
–	168.1	Nonprotonated
–	152.0	Nonprotonated
–	149.5	Nonprotonated
–	139.3	Nonprotonated
–	139.1	Nonprotonated
–	135.3	Nonprotonated
–	135.0	Nonprotonated
7.57	131.82[a]	Methine
7.48	131.78[a]	Methine
7.27	130.9	Methine
7.25	130.8	Methine
7.28	129.6[a,b]	Methine
7.17	129.4[a]	Methine
7.09	127.8	Methine
7.13	127.6	Methine
7.44	125.9	Methine
7.55	121.8	Methine
–	120.9	Nonprotonated
–	120.1	Nonprotonated
–	119.9	Nonprotonated
–	117.0	Nonprotonated
–	95.4	Nonprotonated
–	95.3	Nonprotonated
–	87.0	Nonprotonated
–	86.3	Nonprotonated
–	66.4	Nonprotonated
–	58.8	Nonprotonated
5.42	54.7	Methine
3.79	53.1	Methyl
3.76	52.6	Methyl

TABLE 5.1.3 ^{1}H and ^{13}C NMR Signals of Molecule 5.1 Listed by Group Type—cont'd

^{1}H Signal (ppm)	^{13}C Signal (ppm)	Group Type
4.55	49.7	Methine
2.43	21.81	Methyl
2.38	21.78	Methyl

^{a}*Doubly intense.*
^{b}*Two signals that arise from one ^{13}C site are averaged.*

aligned favorably, we do not expect a gHMBC cross peak between the signals of H25 and C37. Therefore the more downfield aliphatic methine ^{1}H signal at 5.42 ppm is assigned to site 25: $\delta_{H} = 5.42$ ppm and $\delta_{C} = 54.7$ ppm. The more downfield of the two carbonyl ^{13}C signals is assigned to site 35: $\delta_{C} = 169.6$ ppm. We can now assign the other aliphatic methine ^{1}H/^{13}C signal pair to site 28: $\delta_{H} = 4.55$ ppm and $\delta_{C} = 49.7$ ppm. We assign the other carbonyl ^{13}C signal to site 37: $\delta_{C} = 168.1$ ppm.

At this early point in our assignment of NMR signals to molecular sites, we can nevertheless obtain several pieces of circumstantial evidence that support our critical initial assignments.

Our first confirmation stems from the fact that we expect H28, which is bound to a carbon that is part of a three-membered ring, to generate a signal with a chemical shift that is upfield relative to the chemical shift of the H25 signal. Strained rings often feature ^{1}H NMR signals that are upfield (at a lower chemical shift) compared to where we would expect to observe them in a less strained chemical environment.

Our second piece of evidence confirming the assignment of sites 25 versus 28 requires consideration of proximate spins. We expect the signals of H25 and H28 to correlate more or less analogously to the NMR signals of the C29 to C34 aromatic ring. For example, just as H25 couples to C29, so should H28 couple to C34. However, H25 is expected to correlate more strongly in the gHMBC spectrum to ^{13}C signals from the C10 to C15 aromatic ring. Site 28 is not totally isolated from the C10 to C15 aromatic ring as we can see from Fig. 5.1.1, for H28 may generate a $^{3}J_{CH}$ with C15, thereby generating a gHMBC cross peak between signals of H28 and C15. Nonetheless, we expect the H28 signal to share fewer cross peaks with the C10-C15 aromatic ring's ^{13}C signals. The ^{1}H signal at 5.42 ppm that we have assigned to H25 shares gHMBC correlations with six aromatic ^{13}C resonances, while the H28 signal at 4.55 ppm only correlates with four aromatic ^{13}C signals. Our assignment of H25 is therefore consistent with our expectation that the H25 signal will exhibit greater number of correlations to the signals of aromatic ^{13}C's in the gHMBC spectrum than will the signal of H28.

Having confirmed that the NMR signals from sites 25 and 28 are assigned correctly, we are now confident that our carbonyl ^{13}C assignments are also correct. The midfield methyl ^{1}H signal at 3.79 ppm is observed to correlate with the C37 signal at 168.1 ppm (aka 14 ppm) in the gHMBC spectrum. We assign this methoxy ^{1}H signal to site 38: $\delta_{H} = 3.79$ ppm and $\delta_{C} = 53.1$ ppm. The other midfield methyl ^{1}H signal at 3.76 ppm correlates with the C35 signal at 169.6 ppm (aka 15 ppm) in the gHMBC spectrum. We assign the other ^{1}H signal to site 36: $\delta_{H} = 3.76$ ppm and $\delta_{C} = 52.6$ ppm.

Our only two ^{13}C signals from nonprotonated, sp^{3}-hybridized carbon sites are found at 66.4 and 58.8 ppm. These signals are attributed to sites 26 and 27, but as of yet we have not differentiated between them. Because H28 resides on an sp^3-hybridized carbon atom and is two bonds from both C26 and C27, we expect to observe moderate to strong gHMBC cross peaks of nearly equally intensity between the signals of H28 and C26 versus between those of H28 and C27. On the other hand H25, which is three bonds removed from C26 and, due to geometrical constraints, has a dihedral angle in the range of 110–130 degrees, will share only a 2–4 Hz $^3J_{CH}$ with C26. The small $^3J_{CH}$ coupling between H25 and C26 generates a less intense gHMBC cross peak compared to the cross peak intensity we would observe from spins coupled via a $^2J_{CH}$ through an sp^3 carbon center. The H28 signal is found to correlate nearly equally in the gHMBC spectrum with the ^{13}C resonances identified as corresponding to sites 26 and 27, while the H25 signal correlates more strongly with the more downfield (66.4 ppm, not 58.8 ppm) of the two ^{13}C signals from nonprotonated, sp^3-hybridized carbons. Therefore we assign the more upfield of the two ^{13}C signals under consideration to site 26: $\delta_C = 58.8$ ppm and the other ^{13}C signal we assign to site 27: $\delta_C = 66.4$ ppm.

A confirmation of the site 26 versus 27 assignments is obtained by noting a medium intensity gHMBC cross peak between the signal of an aromatic ^1H at 7.25 ppm and the C26 signal at 58.8 ppm. This gHMBC cross peak must be the correlation of the H11 and C26 signals which arises from the $^4J_{CH}$ facilitated by the planarity of the following atoms H11-C11-C10-C15-C26, giving four bonds in the shape of the letter W. H30 is also four bonds from C26, but the bonds between the two spins are not coplanar. The other possible aromatic ^1H's that might couple to C26 (or C27, if our assignment of 26 versus 27 is wrong) are H30 and H33, but again the bond geometry is not coplanar and therefore the $^4J_{CH}$'s are expected to be near zero. We assign site 11 as follows: $\delta_H = 7.25$ ppm and $\delta_C = 130.9$ ppm.

Just as the H11 signal correlates with that of C26 in the gHMBC spectrum, so, too, should the H12 signal correlate with that of C25. Although a t_1 noise ridge is present at the ^1H chemical shift of 7.28 ppm, we observe a somewhat weaker gHMBC cross peak between the aromatic ^1H signal at 7.27 ppm and the C25 signal at 54.7 ppm. We are now able to assign site 12 as follows: $\delta_H = 7.27$ ppm and $\delta_C = 130.8$ ppm.

We now use the COSY spectrum to identify the chemical shifts of the signals of the four aromatic ^1H's of sites 30–33. The two ^1H signals that appear as triplets at 7.13 and 7.09 ppm must be those of H31 and H32, even though we are not able to easily identify their COSY correlation because of its proximity to the diagonal. The triplet ^1H signal at 7.13 ppm correlates with the doublet ^1H signal at 7.44 ppm and the triplet ^1H signal at 7.09 ppm correlates with the doublet ^1H signal at 7.55 ppm. We can therefore order the H30-H33 spin system as 7.55, 7.09, 7.13, 7.44 ppm, but which end is which we have yet to determine.

The well-resolved aromatic ^1H signal (a doublet) at 7.44 ppm shares a gHMBC cross peak with the C28 signal at 49.7 ppm. This aromatic signal is assigned to site 30: $\delta_H = 7.44$ ppm and $\delta_C = 125.9$ ppm. We find an analogous gHMBC cross peak between the aromatic ^1H doublet signal at 7.55 ppm (at the other end of our four-member spin system) and the C25 signal at 54.7 ppm. We assign the 7.55 ppm doublet to site 33: $\delta_H = 7.55$ ppm and $\delta_C = 121.8$ ppm. We now use the order of spins we established previously to write for site 32: $\delta_H = 7.09$ ppm and $\delta_C = 127.8$ ppm and for site 31: $\delta_H = 7.13$ ppm and $\delta_C = 127.6$ ppm. Given that $^4J_{CH}$'s are smaller (<4 Hz) than cis-$^3J_{CH}$'s (~8 Hz), we expect that the two gHMBC cross peaks upon which we base this portion of the assignment are not from correlations between the signals of H30 and C25 and between the signals of H33 and C28.

The H32 signal at 7.09 ppm correlates, as we expect, to the C30 signal at 125.9 ppm, just as the H31 signal at 7.13 ppm correlates with the C33 signal at 121.8 ppm. Both the gHMBC cross peaks correlating the signals of H32 with C30 and H31 with C33 arise from *trans*-$^3J_{CH}$'s. H32 and H31 also are expected to participate in similar *trans*-$^3J_{CH}$ couplings to C34 and C29, respectively. We observe that the H32 signal at 7.09 ppm also correlates strongly in the gHMBC spectrum with the nonprotonated aromatic ^{13}C signal at 149.5 ppm. We assign this ^{13}C signal to site 34: $\delta_C = 149.5$ ppm. Similarly, the H31 signal at 7.13 ppm shares a strong gHMBC cross peak with the nonprotonated aromatic ^{13}C signal at 135.0 ppm. We assign this ^{13}C signal to site 29: $\delta_C = 135.0$ ppm.

We are able to confirm the consistency of our assignments of sites 30 and 33 by noting that the H30 signal at 7.44 ppm correlates in the gHMBC spectrum with the C34 signal at 149.5 ppm, with the C32 signal at 127.8 ppm, and with the C28 signal at 49.7 ppm. H33, whose signal is observed at 7.55 ppm, couples analogously via a *trans*-$^3J_{CH}$ with C29, whose signal is observed at 135.0 ppm, via a *trans*-$^3J_{CH}$ with C31, whose signal is observed at 127.6 ppm, and via a *cis*-$^3J_{CH}$ with C25, whose signal is observed at 54.7 ppm. Recall that we do not expect H30 to couple strongly with C29 because C30 is sp^2-hybridized. In the same manner, H33 is not expected to couple strongly to C34 because C33 is sp^2-hybridized.

The H28 signal at 4.55 ppm correlates with three already assigned ^{13}C signals, that of C34 at 149.5 ppm, that of C29 at 135.0 ppm, and that of C30 at 125.9 ppm. The H28 signal also correlates in the gHMBC spectrum with the unassigned ^{13}C signal at 135.3 ppm. Only C15 is within three bonds of H28 ($^4J_{CH}$'s are not expected because of the lack of planarity between the C28-H28 bond and the potential coupling partners C10 and C14), and so we write for site 15: $\delta_C = 135.3$ ppm. Notice that the cross peaks between the signals of H28 and C15 (4.55 ppm, 135.3 ppm) and between the signals of H25 and C15 (5.42 ppm, 135.3 ppm) are overlapped with the cross peaks between the signals of H28 and C29 and also between those of H25 and C29, respectively. We are still confident that the gHMBC cross peaks we observe involving the C15 signal are real (and not a misattribution of the edges of the high intensity cross peaks between the signals of H28 and C29, and between those of H25 and C29), because, if we look below the location of these cross peaks (f_1 is horizontal), we can observe for example that the cross peak between the H31 signal at 7.13 ppm and the C29 signal at 135.0 ppm is significantly narrower, suggesting that the ^{13}C (f_1) width of the cross peak between the signals of H29 and C29 is not sufficient to account for the gHMBC intensity we attribute to the H28-C15 interaction.

The H25 signal at 5.42 ppm is observed in the gHMBC spectrum to correlate with six aromatic ^{13}C's. The four assigned aromatic ^{13}C signals to which the H25 signal correlates in the gHMBC spectrum are those of C34 at 149.5 ppm, C15 at 135.3 ppm, C29 at 135.0 ppm, and C33 at 121.8 ppm. The other two gHMBC correlations between the H25 signal and aromatic ^{13}C signals are to the unassigned signals at 152.0 and 117.0 ppm.

The H25 signal at 5.42 ppm is expected to share a stronger gHMBC cross peak with the C14 signal than with the C13 or C15 signal at 135.3 ppm because the C25-H25 bond is not expected to lie in the plane of the C10-C15 aromatic ring. The H25 signal correlates more strongly with the ^{13}C signal at 152.0 ppm compared with that at 117.0 ppm, and so we write for site 14: $\delta_C = 152.0$ ppm. We assign the other ^{13}C signal that correlates in the gHMBC spectrum with H25 to site 13: $\delta_C = 117.0$ ppm.

Although overlap is significant, we are still able to observe a gHMBC correlation between the H12 signal at 7.27 ppm and the newly assigned C14 signal at 152.0 ppm. The H12 signal at

7.27 ppm also correlates in the gHMBC spectrum with the nonprotonated aromatic ^{13}C signal at 120.9 ppm. We assign this ^{13}C signal to site 10: $\delta_C = 120.9$ ppm.

The H11 signal at 7.25 ppm is observed to correlate in the gHMBC spectrum with the C15 signal at 135.3 ppm and with the C13 signal at 117.0 ppm, confirming the assignments we have made up to this point.

The H11 signal at 7.25 ppm is observed to correlate in the gHMBC spectrum with the midfield nonprotonated ^{13}C signal at 86.3 ppm. We assign this ^{13}C signal to site 9: $\delta_C = 86.3$ ppm.

The H12 signal at 7.27 ppm shares a gHMBC cross peak with the ^{13}C signal at 87.0 ppm. We assign this ^{13}C signal to site 16: $\delta_C = 87.0$ ppm. The other two sp-hybridized ^{13}C signals are found at 95 ppm, but we cannot yet assign these resonances.

Quite incredibly, the site 1 and 24 methyl ^{1}H signals correlate with the signals of C9 and C16 in the gHMBC spectrum. The gHMBC cross peaks must arise from a $^7J_{CH}$ (!). The methyl ^{1}H signal at 2.38 ppm shares a gHMBC cross peak with the C9 signal at 86.3 ppm. We assign this ^{1}H signal to site 1: $\delta_H = 2.38$ ppm and $\delta_C = 21.78$ ppm. The methyl ^{1}H signal at 2.43 ppm similarly correlates with the C16 signal at 87.0 ppm. We assign the methyl ^{1}H signal to site 24: $\delta_H = 2.43$ ppm and $\delta_C = 21.81$ ppm. Molecular rigidity and conjugation facilitate the long-range coupling.

The H1 signal at 2.38 ppm also correlates strongly in the gHMBC spectrum with a nonprotonated ^{13}C signal at 139.1 ppm and more weakly to a nonprotonated ^{13}C signal at 119.9 ppm. We assign the strongly correlated ^{13}C signal to site 2: $\delta_C = 139.1$ ppm and the weakly correlated ^{13}C signal to site 5: $\delta_C = 119.9$ ppm.

The H24 signal at 2.43 ppm shares an analogously strong gHMBC cross peak with the nonprotonated ^{13}C resonance at 139.3 ppm. We assign this ^{13}C signal to site 21: $\delta_C = 139.3$ ppm. The H24 signal also shares a weaker gHMBC cross peak with a second nonprotonated aromatic ^{13}C signal at 120.1 ppm. We assign this second ^{13}C signal to site 18: $\delta_C = 120.1$ ppm.

The ^{1}H signals from sites 1 and 24 share weak cross peaks with the signals of ^{1}H's on the C2-C7 and C18-C23 aromatic rings, respectively. H1 is expected to couple more strongly to the H3/7 pair than to the H4/6 pair. In the COSY spectrum the H1 signal at 2.38 ppm correlates moderately with the ^{1}H signal at 7.17 ppm and more weakly with the ^{1}H signal at 7.48 ppm. We assign the moderate correlation to the H1-H3/7 interaction, writing for sites 3/7: $\delta_H = 7.17$ ppm and $\delta_C = 129.4$ ppm. We attribute the weaker correlation to the $^4J_{HH}$ between H1 and H4/H6, writing for sites 4/6: $\delta_H = 7.48$ ppm and $\delta_C = 131.78$ ppm. The H24 signals at 2.43 ppm share a moderate COSY correlation with the doubly intense aromatic ^{1}H signals at 7.28 ppm. We assign the 7.28 ppm signal to sites 20/22: $\delta_H = 7.28$ ppm and $\delta_C = 129.6$ ppm. The H24 signal at 2.43 ppm also correlates with aromatic ^{1}H signal at 7.57 ppm. We assign this last ^{1}H signal to sites 19/23: $\delta_H = 7.57$ ppm and $\delta_C = 131.82$ ppm.

Returning to the gHMBC spectrum, we observe that the H1 signal at 2.38 ppm shares a discernible cross peak with the nonprotonated ^{13}C signal at 95.4 ppm, which we assign to site 8: $\delta_C = 95.4$ ppm. The H24 signal at 2.43 ppm correlates with the nonprotonated ^{13}C signal at 95.3 ppm, and so we write for site 17: $\delta_C = 95.3$ ppm.

We are able to confirm much of the consistency of our assignments using the gHMBC spectrum. For instance, the H4/6 signal at 7.48 ppm shares a gHMBC cross peak with the C2 signal at 139.1 ppm, and the H19/23 signal at 7.57 ppm correlates in the gHMBC with the C21 signal at 139.3 ppm.

5.2 A PAIR OF 27-CARBON EPIMERS (MOLECULES 5.2.1 AND 5.2.2)

5.2.1 Molecule 5.2.1 in Benzene-d_6

The IUPAC name for molecule 5.2.1 is (2S,3S,4R,6S)-2-allyl-6-(3-benzyloxy)propyl)-4-(4-methoxybenzyloxy)-3-methyltetrahydro-2H-pyran-3-ol, which when spoken does not freely flow off of the tongue. Samples of molecules 5.2.1 and 5.2.2 were provided to the author by Dr. Denise Colby, working in the laboratory of Dr. Timothy Jamison. Dr. Colby's sample numbers for these two molecules were dc-1-105-2 (molecule 5.2.1) and dc-2-56. The empirical formula for both molecules is $C_{27}H_{36}O_5$. The structure of molecule 5.2.1 appears in Fig. 5.2.1.1. The sole difference between the two molecules is that the stereochemistry of site 2 is inverted for molecule 5.2.2 compared with 5.2.1. That is, site 2 is an S-stereocenter in molecule 5.2.1 and an R-stereocenter in molecule 5.2.2, making the two molecules an epimeric pair of molecules.

The conventions established through Cahn-Ingold-Prelog prioritization allows us to decide whether each stereocenter is, for planar systems, *cis* versus *trans*, and for a 3-D systems, right-handed versus left-handed. Planar systems can be classified as having their highest priority groups *cis* are said to be Z (German for *zusammen*, or together), while planar systems with their highest priority groups *trans* are said to be E (German for *entgegen*, or opposed). It is unfortunate that the physical appearance of the letter Z connotes opposition insofar as the top and bottom of the letter Z are on opposite sides of the diagonal making up the middle of the letter. Meanwhile, the physical appearance of the letter E suggests a grouping on the same side insofar as the three horizontal lines in the letter E all reside on the same side of the letter E's vertical line. The trick is to remember that the connotations are wrong. In our efforts to describe 3-D molecules, we assign the letters R and S. We assign the letter R to right-handed stereocenters. The choice of the letter R derives from the Latin word *rectus* in the sense of right-handed and also *correct*. If our stereocenter is found to be left-handed, we assign it the letter S, linking this descriptor etymologically to the Latin word *sinister*. The Romans apparently believed that left-handed people were inherently evil, a belief to which the left-handed author does not subscribe—tolerance of left-handedness is a surprisingly recent event. To wit, the author's left-handed paternal grandfather was beaten on the left hand in school in Scituate, Massachusetts, United States c.1901 for writing with his left land, and so he wrote with his

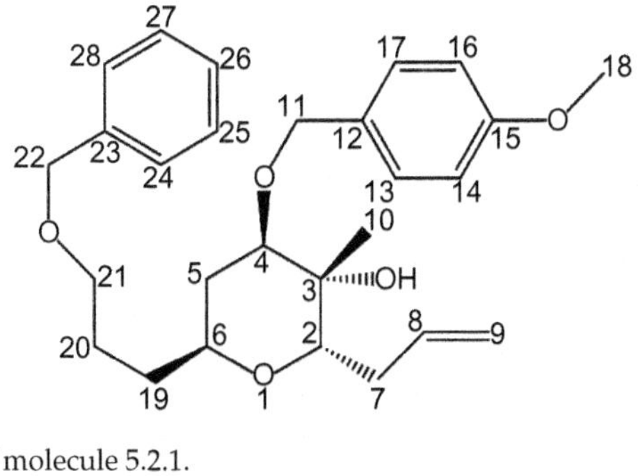

FIG. 5.2.1.1   The structure of molecule 5.2.1.

right. As a result of being forced to use his nondominant hand, his handwriting was terrible. Presumably those reading this are scientists and appreciate that nature is replete with examples of right-handedness and left-handedness—sometimes in equal proportion (e.g., a racemic mixture) and sometimes not. The dominance of a given handedness does not confer divine intent. In this case, we can laugh about the inherent social bias introduced by using R and S stereocenter descriptors, but at the same time we might resolve to be vigilant to the subtle and pernicious effects introduced through choice of language.

We will assign the spectra of molecule 5.2.1 and then we will easily be able to assign signals to sites in molecule 5.2.2, except that we will note more profound overlap in the aromatic ^{13}C signals of molecule 5.2.2. Because the aromatic ^{13}C signal overlap in the spectra of molecule 5.2.2 is generated by a portion of the molecule well removed from the site of epimerization, our discussions of the site inversion and its spectral consequences (specifically the gHMBC interaction between the signals of H2 and C10) are not impacted (other than to note that the overlap occurs in one molecule but not the other due solely to site 2 inversion). This teaches us that molecular conformation likely depends on many competing and sometimes subtle factors, and how solvent molecules are able to pack around a given solute molecule with multiple stereocenters must surely affect local chemical environment throughout a solute molecule.

Molecule 5.2.1 has two aromatic rings, an allyl (methylene plus terminal vinyl) group, four ether linkages, and a hydroxyl group. The molecule was dissolved in benzene-d_6 and a single sample was used to collect all spectra. The 1-D ^{1}H NMR spectrum of molecule 5.2.1 appears in Fig. 5.2.1.2. We note the excellent signal dispersion that exists in the spectrum,

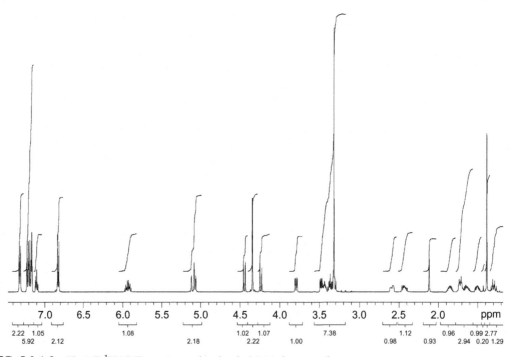

FIG. 5.2.1.2   The 1-D ^{1}H NMR spectrum of molecule 5.2.1 in benzene-d_6.

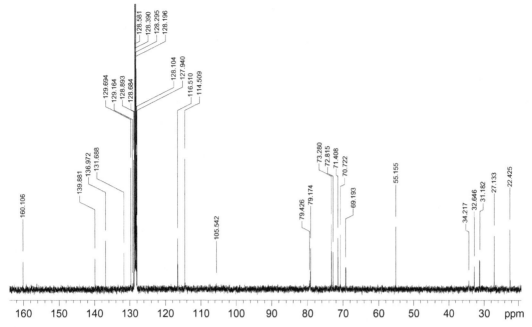

FIG. 5.2.1.3   The 1-D ^{13}C NMR spectrum of molecule 5.2.1 in benzene-d_6.

boding well for the pairing of signals and sites. We also observe that our integrals are not as close to integer values as we might like, but must acknowledge that when we have one real-world sample and spectrometer time is in demand, we may not have the luxury of collecting a 1-D ^{1}H NMR spectrum whose observed spins are completely relaxed. Information found in the HSQC spectrum will complement the ^{1}H integral information sufficiently to allow unequivocal assignments. The 1-D ^{13}C NMR spectrum of molecule 5.2.1 appears in Fig. 5.2.1.3. We can readily pick out C15 as generating the most downfield ^{13}C signal owing to the combination of aromaticity and proximity to oxygen. We note a lack of correspondence between the number of carbon atoms in our molecule and the number of ^{13}C signals (two too many) found by our pick picking algorithm (even after accounting for the 1:1:1 solvent triplet at 128.39 ppm), especially in the aromatic chemical shift region. The overabundance of ^{13}C signals found by the peak picker occurs because the peak picking threshold had to be set at a very low value to successfully register the ^{13}C signals at 34.2 and 79.4 ppm, which are broad, possibly the result of a slow conformational rearrangement.

Beside the 1:1:1 triplet signal arising from deuterated ^{13}C's, we also may be observing the ^{13}C signal from the protonated ^{13}C in the benzene-d_5,h_1 species. ^{13}C's alpha to protonated carbons may also generate a weaker 1:1:1 triplet signal whose total (fully relaxed) integrated area would be twice that of the protonated ^{13}C signal, since for every protonated carbon there will be two neighboring deuterated carbon sites with this anomalously protonated carbon one bond removed. Fig. 5.2.1.4 shows the 2-D ^{1}H-^{1}H COSY NMR spectrum of molecule 5.2.1. The COSY spectrum has many well-resolved cross peaks, although the aromatic chemical shift region is crowded and also complicated by the benzene-d_5,h_1 signal at 7.16 ppm. The 2-D ^{1}H-^{13}C HSQC NMR spectrum appears in Fig. 5.2.1.5 and contains the information we require to

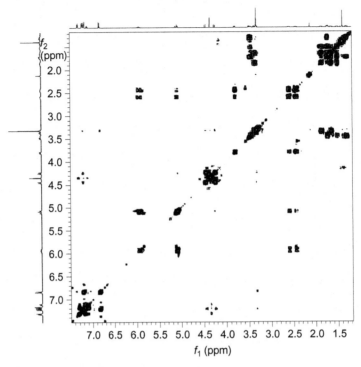

FIG. 5.2.1.4   The 2-D ^{1}H-^{1}H COSY NMR spectrum of molecule 5.2.1 in benzene-d_6.

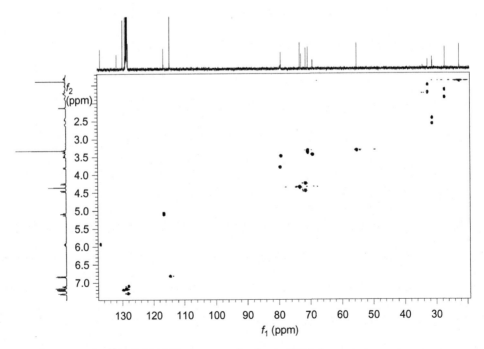

FIG. 5.2.1.5   The 2-D ^{1}H-^{13}C HSQC NMR spectrum of molecule 5.2.1 in benzene-d_6.

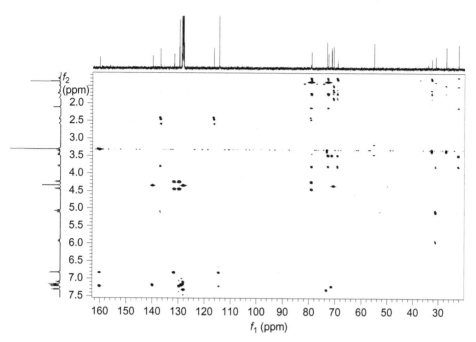

FIG. 5.2.1.6   The 2-D ^{1}H-^{13}C gHMBC NMR spectrum of molecule 5.2.1 in benzene-d_6.

differentiate the NMR signals of the solute and solvent. In the upfield of the HSQC spectrum we are missing a pair of methylene cross peaks, which in retrospect are found to be missing because the contour drawing threshold of the HSQC spectrum was set too high. In particular, the H5$_{ax}$ (axial ^{1}H on site 5) couples strongly to two vicinal axial ^{1}H's (H4 and H6) and so its signal is spread out into an approximate quartet because of the strong geminal coupling to the H5$_{eq}$ (equatorial ^{1}H on site 5) as well as the two trans-$^3J_{HH}$'s to H4 and H6. The 2-D ^{1}H-^{13}C gHMBC NMR spectrum of molecule 5.2.1 is shown in Fig. 5.2.1.6. This single spectrum contains perhaps half of all the information required to achieve the complete and unambiguous pairing of signals and molecular sites.

As is our custom, we examine the structure of molecule 5.2.1 and predict the multiplicities of the ^{1}H signals that we anticipate observing, placing our predictions into Table 5.2.1.1. We assume that the aliphatic ring encompassing sites 1–6 is in the chair conformation with the OC11 and C19 groups equatorial, since these are the bulkiest pendant groups that presumably provide the lowest energy when occupying an equatorial position. The assumption of the chair conformation for the saturated ring may not be a good one, as C7 becomes axial if C9 and OC11 are equatorial. The actual conformation may be a skew boat for this epimer. If we invert site 2, the ring is more likely to adopt the chair conformation.

We place into Table 5.2.1.2 our accounting of the numbered sites and heteroatom-bound ^{1}H's in molecule 5.2.1. Table 5.2.1.3 contains the chemical shifts of the ^{1}H and ^{13}C NMR signals from molecule 5.2.1.

We have a number of entry points from which to choose. We can readily identify the C15 signal by its chemical shift near 150 ppm, we can identify the H26 signal as the only aromatic

TABLE 5.2.1.1   Predicted Multiplicities for the ^{1}H's of Molecule 5.2.1

Site in Molecule	Expected Multiplicity
2	d^2
3 (OH)	s
4	d^2
5	q, dt
6	d^4
7	$2 \times d^3$
8	d^4
9	$2 \times d^2$
10	s
11	$2 \times d$
13/17	d^a
14/16	d^a
18	s
19	$2 \times d^4$
20	$2 \times d^5$
21	$2 \times d^3$
22	$2 \times d$
24/28	d^a
25/27	t^a
26	t

^{a}*Double intensity.*

TABLE 5.2.1.2   Numbered Sites of Molecule 5.2.1

Type of Site	Site Number
CH_3 (methyl)	10, 18
CH_2 (methylene)	5, 7, 9, 11, 19, 20, 21, 22
CH (methine)	2, 4, 6, 8, 13/17, 14/16, 24/28, 25/27, 26
C_{np} (nonprotonated)	3, 12, 15, 23
Numbered heteronucleus	1 (O)
Protonated heteronucleus	3OH

TABLE 5.2.1.3 ^{1}H and ^{13}C NMR Signals of Molecule 5.2.1 Listed by Group Type

^{1}H Signal (ppm)	^{13}C Signal (ppm)	Group Type
–	160.1	Nonprotonated
–	139.9	Nonprotonated
5.93	137.0	Methine
–	131.7	Nonprotonated
7.22	129.7^a	Methine
7.17	128.9^{a,b}	Methine
7.31	128.2^{a,b}	Methine
7.10	127.9	Methine
5.08 & 5.10	116.5	Methylene
6.83	114.5^a	Methine
3.79	79.4	Methine
3.48	79.2	Methine
4.34 & 4.35	73.3	Methylene
–	72.8	Nonprotonated
4.24 & 4.44	71.4	Methylene
3.33 & 3.37	70.7	Methylene
3.44	69.2	Methine
3.32	55.2	Methyl
1.30^c & 1.72	34.2	Methylene
1.51 & 1.71	32.6	Methylene
2.43 & 2.58	31.2	Methylene
1.65 & 1.86	27.1	Methylene
1.38	22.4	Methyl
2.12	–	Hydroxyl

^{a}Doubly intense.
^{b}Extra ^{13}C signals observed near this signal.
^{c}The HSQC cross peak involving this ^{1}H signal is not observed.

^{1}H resonance that integrates to one hydrogen equivalent instead of two, and we can easily choose either methyl group as well, since site 18 is alpha to oxygen (and hence its signal is more downfield) and site 10 is not.

We instead begin with the vinyl group, part of the allyl group, at sites 8 and 9. The site 9 methylene group is easily identified because it is the only sp^2-hybridized carbon site with two distinct hydrogens. The most downfield methylene group ^{1}H signals are found at the chemical shifts of 5.08 & 5.10 ppm. Notice the profound nonfirst-order skewing of the intensity away from the outer the legs of the H9 signal doublets in the 1-D ^{1}H NMR spectrum. If there is any doubt, we can observe these 5.1 ppm ^{1}H signals coupling with another downfield ^{1}H signal, at 5.93 ppm in the COSY spectrum. We assign site 9 as follows: $\delta_H = 5.08$ & 5.10 ppm and $\delta_C = 116.5$ ppm. Site 8 is assigned to the ^{1}H signal at 5.93 ppm, and so we write for site 8: $\delta_H = 5.93$ ppm and $\delta_C = 137.0$ ppm.

The H8 signal at 5.93 ppm correlates with the methylene H7 signals. In the COSY spectrum we observe that the H8 signal at 5.93 ppm, besides correlating with the H9 signals, also correlates with two methylene ^{1}H signals at 2.43 & 2.58 ppm. We assign these methylene ^{1}H signals to site 7: $\delta_H = 2.43$ & 2.58 ppm and $\delta_C = 31.2$ ppm.

Moving along the spin system into the saturated six-membered ring, we observe that the H7 signals at 2.43 & 2.58 ppm correlate in the COSY spectrum with a midfield methine doublet of doublets signal at 3.79 ppm. We assign this methine ^{1}H signal at 3.79 ppm to site 2: $\delta_H = 3.79$ ppm and $\delta_C = 79.4$ ppm.

Having reached the end of this ^{1}H spin system, we must switch to the gHMBC spectrum to continue or find another entry point, of which there are many. Economy is achieved by doing both. To do this, we find a nearby methyl group as a second entry point. We can easily distinguish between the NMR signals of the two methyl groups: The site 10 methyl group is bound to carbon, while the site 18 methyl is bound to oxygen. Site 18 generates a signal that appears downfield (at higher ppm values) from that of site 10, and so we can identify the site 10 NMR signals as follows: $\delta_H = 1.38$ ppm and $\delta_C = 22.4$ ppm. We write for site 18: $\delta_H = 3.32$ ppm and $\delta_C = 55.2$ ppm.

Both the signals of the H10's at 1.38 ppm and that of H2 at 3.79 ppm are found to share gHMBC correlations with the nonprotonated midfield ^{13}C signal at 72.8 ppm. Since site 3 is the only nonprotonated sp^3-hybridized carbon, we assign this ^{13}C signal to site 3: $\delta_C = 72.8$ ppm. The signal at 2.12 ppm arising from a ^{1}H on a heteroatom also shares a gHMBC cross peak with the C3 signal at 72.8 ppm, and so we write for site 3 (OH): $\delta_H = 2.12$ ppm. The main problem with identifying hydroxyl (and amino) ^{1}H signals is that in many cases other impurity signals may interfere with our ability to unambiguously identify the signal of a heteroatom-bound ^{1}H. The gHMBC correlation between the signals of the hydroxyl ^{1}H and ^{13}C's found two bonds away removes any doubt as to the actual identity of the hydroxyl ^{1}H signal at 2.12 ppm.

Interestingly, the site 3 hydroxyl ^{1}H signal also shares a gHMBC cross peak with a midfield methine ^{13}C signal at 79.2 ppm. We conclude that this methine ^{13}C signal belongs to site 4 (as opposed to site 6), writing for site 4: $\delta_H = 3.48$ ppm and $\delta_C = 79.2$ ppm. We observe the gHMBC cross peak between the signals of H2 and C4 which is consistent with our assignment (but H2 is expected to couple just as well to C6, so this particular piece of information does not help us to differentiate between the signals of sites 4 and 6). Furthermore, because the ^{13}C NMR signals of C2 and C4 are within 0.2 ppm of one another, we are not at this point able to confirm our assignment using a correlation between the signals of H10 and C4, because the H10 signal surely correlates with the C2 signal just as well. Nonetheless, we can argue that the hydroxyl ^{1}H on site 3 is more likely to couple to C4, which is three bonds distant, than to C6, which is five bonds distant.

The H2 signal at 3.79 ppm shares a gHMBC cross peak with another (i.e., besides C4) methine ^{13}C signal at 69.2 ppm. We assign this ^{13}C signal to site 6: $\delta_H = 3.44$ ppm and $\delta_C = 69.2$ ppm.

We expect both the signals of H4 at 3.48 ppm and H6 at 3.44 ppm to correlate in the gHMBC spectrum with the C5 signal, but we do not see any cross peaks that might reasonably qualify as resulting from this interaction. Recalling the broadening of one of the upfield ^{13}C signals (in the context of the discussion of why the peak picking algorithm found too many ^{13}C signals) that required us to lower the peak picking threshold, we might suppose that conformational exchange occurring in the site 1–6 ring may contribute to the broadening of one of more of the NMR signals associated with this ring.

Two upfield ^{1}H signals at 1.30 and 1.72 ppm share gHMBC cross peaks with the signals of C4 at 79.2 ppm, with C3 at 72.8 ppm, and C6 at 69.2 ppm. We assign these upfield signals to the methylene ^{1}H's of site 5: $\delta_H = 1.30$ & 1.72 ppm and $\delta_C = 34.2$ ppm. If we examine the HSQC spectrum closely, we observe that the more downfield H5 resonance at 1.72 ppm shares a weak HSQC correlation with one of the weakest observed ^{13}C signals at 34.2 ppm. Recall that in six- membered rings in the chair conformation, the NMR signal of an equatorial ^{1}H is often found downfield (at a higher ppm value) from that of its axial counterpart. A possible explanation for the failure to observe both HSQC correlations for the site 5 spins is that the multiple strong J-couplings involving the axial ^{1}H on site 5, combined with the broadness of the C5 resonance owing to conformational variation on the NMR time scale serve to spread out the cross peak between the signals of H5$_{ax}$ and C5 so thinly that its height fails to exceed the 2-D contour plotting threshold in the HSQC spectrum.

The H18 signal at 3.32 ppm shares a strong gHMBC cross peak with the ^{13}C signal at 160.1 ppm. With the certainty of knowing that only C15 is aromatic and alpha to oxygen, we assign this ^{13}C signal to site 15: $\delta_C = 160.1$ ppm.

Aromatic ^{1}H's couple more strongly to ^{13}C's that are three than to ^{13}C's that are two bonds distant, we therefore expect the H13/17's will couple more strongly with C15 than will the H14/16's. We observe a strong gHMBC cross peak between the aromatic ^{1}H signal at 7.21 ppm and the C15 signal at 160.1 ppm, while in contrast we observe a medium intensity gHMBC cross peak between the aromatic ^{1}H signal at 6.83 ppm and the C15 signal at 160.1 ppm. We assign the ^{1}H signal at 7.21 ppm, which is more downfield and more strongly correlated to the C15 signal, to sites 13/17: $\delta_H = 7.21$ ppm and $\delta_C = 129.7$ ppm. We assign the other aromatic ^{1}H signal to sites 14/16: $\delta_H = 6.83$ ppm and $\delta_C = 114.5$ ppm.

The relative chemical shifts of sites 13/17 versus 14/16 are consistent with our understanding that electron density donation from the lone pairs of the methoxy oxygen atom on site 15 will shift the signal of the spins *ortho* to site 15 (sites 14/16) upfield relative to the signals of their *meta* counterparts (sites 13/17). This is why the site 13/17 NMR signals are downfield relative to the site 14/16 signals.

The signals of the H14/16's are also expected to correlate strongly with the C12 signal in the gHMBC spectrum. We observe a gHMBC cross peak between the H14/16 signals at 6.83 ppm and a nonprotonated aromatic ^{13}C signal at 131.7 ppm. We assign this ^{13}C signal to site 12: $\delta_C = 131.7$ ppm.

A midfield methylene ^{1}H signal pair at 4.24 & 4.44 ppm is found to correlate with the C12 signal at 131.7 ppm, with the C13/17 signal at 129.7 ppm, and with the C4 signal at 79.2 ppm. We assign this pair of methylene ^{1}H signals to site 11: $\delta_H = 4.24$ & 4.44 ppm and $\delta_C = 71.4$ ppm.

We must still identify the signals of sites 21 and 22, both of which are methylene groups adjacent to oxygen, as well as those of the methylenes 19 and 20. We will begin afresh at another site.

The site 26 ^{1}H signal is easily located at 7.10 ppm in the 1-D ^{1}H NMR spectrum, for it is the only aromatic ^{1}H signal that integrates to a value near one. We write for site 26: $\delta_H = 7.10$ ppm and $\delta_C = 127.9$ ppm.

The H26 signal at 7.10 ppm shares a COSY cross peak with the aromatic ^{1}H signal at 7.17 ppm, which is the more upfield of the two aromatic ^{1}H signals whose signals span the chemical shift ranges of 7.15 to 7.24 ppm. We assign this 7.17 ppm ^{1}H signal to sites 25/27: $\delta_H = 7.17$ ppm and $\delta_C = 128.9$ ppm.

The other aromatic ^{1}H/^{13}C signal pair is assigned to sites 24/28: $\delta_H = 7.31$ ppm and $\delta_C = 128.2$ ppm, as this is the last downfield ^{1}H/^{13}C signal pair to be assigned. We confirm that the H24/28 signals at 7.31 ppm share a gHMBC cross peak with the C26 signal at 127.9 ppm, although the gHMBC cross peaks between the signals of H24 and C28 and also between those of H28 and C24 are found nearby at 128.2 ppm, partially overlapping the gHMBC cross peak of interest.

The H25/27 signal at 7.17 ppm shares a strong gHMBC cross peak with a nonprotonated aromatic ^{13}C signal at 139.3 ppm. We assign this last aromatic ^{13}C signal to site 23: $\delta_C = 139.3$ ppm.

The H24/28 signal at 7.31 ppm is observed to correlate with a midfield methylene ^{13}C signal at 73.3 ppm. We assign this ^{13}C signal to site 22: $\delta_H = 4.34$ & 4.35 ppm and $\delta_C = 73.3$ ppm. Confirming this assignment, we observe that the H22 signals at 4.34 & 4.35 ppm correlate in the gHMBC spectrum with the C23 signal at 139.3 ppm and with the C24/28 signal at 128.2 ppm (a correlation with the C26 signal at 127.9 ppm is unlikely).

The H22 signals at 4.34 & 4.35 ppm also share a gHMBC cross peak with a midfield methylene ^{13}C signal at 70.7 ppm. We assign this ^{13}C signal to site 21: $\delta_H = 3.33$ & 3.37 ppm and $\delta_C = 70.7$ ppm.

The H21 signals at 3.33 & 3.37 ppm share COSY cross peaks with a pair of methylene ^{1}H signals at 1.65 & 1.86 ppm. We assign the latter pair of methylene ^{1}H signals to site 20: $\delta_H = 1.65$ & 1.86 ppm and $\delta_C = 27.1$ ppm. The H21 signals at 3.33 & 3.37 ppm share a confirming gHMBC cross peak with the C20 signal at 27.1 ppm.

The H21 signals at 3.33 & 3.37 ppm also correlate in gHMBC spectrum with a second methylene ^{13}C signal at 32.6 ppm. This ^{13}C signal also correlates in the gHMBC spectrum with both the H5 signals at 1.30 & 1.72 ppm. Only C19 is positioned to couple to the H21's and the H5's, and so we write for site 19: $\delta_H = 1.51$ & 1.71 ppm and $\delta_C = 32.6$ ppm.

Interestingly, we fail to observe gHMBC cross peaks between the H20 signals at 1.65 & 1.86 ppm and the C19 signal at 32.6 ppm. We also do not observe a gHMBC correlation between the H6 signal at 3.44 ppm and the C19 signal at 32.6 ppm. Only the more downfield of the two H20 signals (at 1.86 ppm) is found to correlate with the C19 signal at 32.6 ppm in the gHMBC spectrum. Why the H21 signals appear to correlate more strongly than the H20 signals with that of C19 is a point to ponder. It may be that only one of the H21's is coupled strongly to C19, because the H21 signals lie very close to one another on the ^{1}H chemical shift axis. The one, well-aligned H21 may be the sole source of the relatively intense gHMBC cross peak between the signals of H21 and C19.

5.2.2 Molecule 5.2.2 in Benzene-d_6

Molecule 5.2.2 is identical to molecule 5.2.1 except it has an R-stereocenter at site 2. The IUPAC name for molecule 5.2.2 is (2R,3S,4R,6S)-2-allyl-6-(3-benzyloxy)propyl)-4-(4-methoxybenzyloxy)-3-methyltetrahydro-2H-pyran-3-ol. A sample of this molecule was furnished to the author by Drs. Denise Colby and Timothy Jamison. Dr. Colby's sample number for this molecule was dc-2-56. The empirical formula for molecule 5.2.2 is $C_{27}H_{36}O_5$.

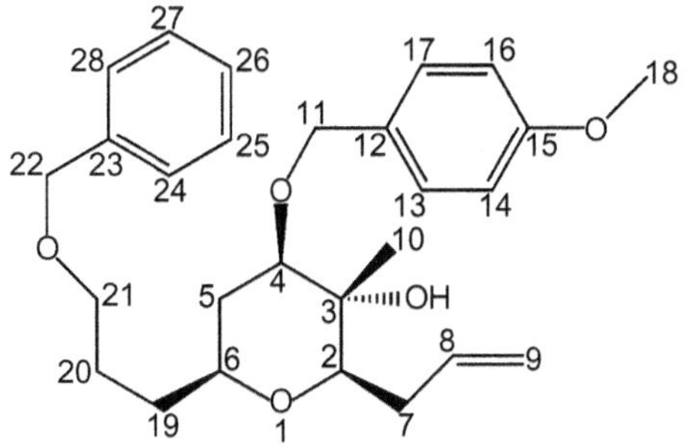

FIG. 5.2.2.1   The structure of molecule 5.2.2.

The structure of molecule 5.2.2 appears in Fig. 5.2.2.1. Note that the structures of molecules 5.2.1 and 5.2.2 differ only at site 2 which is an *S*-stereocenter in molecule 5.2.1 and an *R*-stereocenter in molecule 5.2.2. These molecules are, by definition, epimers.

Molecule 5.2.2 has two aromatic rings, a six-membered ether ring, and three other ether linkages, including a methoxy group, the other methyl residing on an aliphatic carbon. An allyl group is pendant to the saturated six-membered ring. A single hydroxyl group accounts for the fifth oxygen atom. Sites 1–6 comprise the six-membered saturated ring that we are confident adopts the chair conformation. The lowest energy chair conformation of the ring positions the relatively bulky C7, OC11, and C19 in equatorial positions, with the energetically less significant three methine ^{1}H's of sites 2, 4, and 6 being relegated to axial positions. Notice that placement of C7, OC11, and C19 into equatorial positions also forces the site 10 methyl group into an axial position. The axial H2 will share a strong *trans*-$^3J_{CH}$ with C10 as a result of the lowest energy chair conformation. With our understanding of the ^3J Karplus relationship that describes how coupling varies as a function of bond vectors whose origins share opposite ends of a common bond *and* our prediction that C2-H2 bond and the C3-C10 bond will form a 180 degrees dihedral angle, we predict that the intensity of the gHMBC cross peak between the signals of H2 and C10 for molecule 5.2.2 will be stronger than that for molecule 5.2.1.

Aided by our earlier assignment of the observed NMR signals of molecule 5.2.1 to its numbered sites, we are able to very quickly assign the molecule, having only to account for a few small shift perturbations involving sites 4 versus 6, because although the ^{13}C shifts of C7 and C19 are similar and change positions upon site 2 inversion, the ^{1}H chemical shifts of the H7's versus those of the H19's are still significant and their approximate positions along the chemical shift axis are *not* switched. We will come back to this point after all shifts have been assigned. To dispense with the elucidation of how the ^{1}H and ^{13}C NMR signals of molecule 5.2.2 are assigned (intellectually it covers little new ground), the reader can skip ahead to Section 5.2.3 where the epimeric consequences of site 2 inversion are discussed.

The 1-D ^{1}H NMR spectrum of molecule 5.2.2 in benzene-d_6 is shown in Fig. 5.2.2.2. The 1-D ^{13}C NMR spectrum of molecule 5.2.2 appears in Fig. 5.2.2.3. None of the ^{13}C signals of molecule 5.2.2 are as broad as some of the ^{13}C signals we observed previously in our treatment of molecule 5.2.1. The 2-D ^{1}H-^{1}H COSY NMR spectrum of molecule 5.2.2 appears in Fig. 5.2.2.4, the 2-D ^{1}H-^{13}C HSQC NMR spectrum is shown in Fig. 5.2.2.5, and the 2-D ^{1}H-^{13}C gHMBC NMR spectrum is found in Fig. 5.2.2.6. The gHMBC spectrum of molecule 5.2.2 features more cross peaks than that of molecule 5.2.1. The narrowness of all ^{13}C signals of molecule 5.2.2

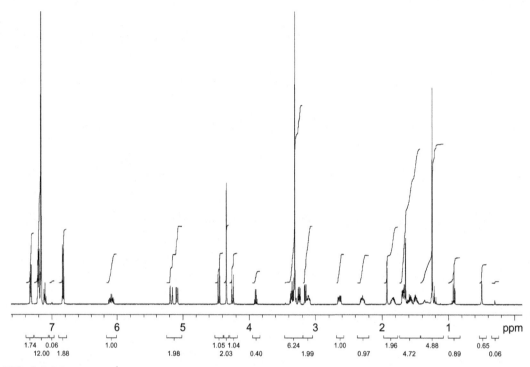

FIG. 5.2.2.2   The 1-D ^{1}H NMR spectrum of molecule 5.2.2 in benzene-d_6.

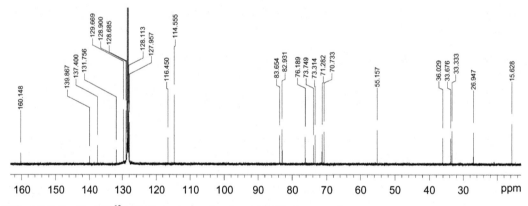

FIG. 5.2.2.3   The 1-D ^{13}C NMR spectrum of molecule 5.2.2 in benzene-d_6.

and the increased prevalence of gHMBC correlations compared with the gHMBC spectrum of molecule 5.2.1 suggest that molecule 5.2.2 is less fluxional than molecule 5.2.1.

Table 5.2.2.1 contains the predicted ^{1}H signal multiplicities we expect for molecule 5.2.2. Table 5.2.2.2 accounts for the various numbered molecular sites and also lists the expected ^{1}H signal for the site 3 hydroxyl group. Table 5.2.2.3 lists all the ^{1}H and ^{13}C NMR signals observed for molecule 5.2.2.

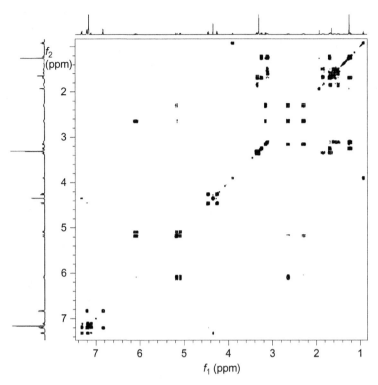

FIG. 5.2.2.4   The 2-D ^{1}H-^{1}H COSY NMR spectrum of molecule 5.2.2 in benzene-d_6.

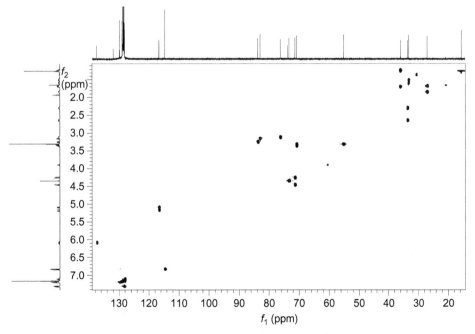

FIG. 5.2.2.5   The 2-D ^{1}H-^{13}C HSQC NMR spectrum of molecule 5.2.2 in benzene-d_6.

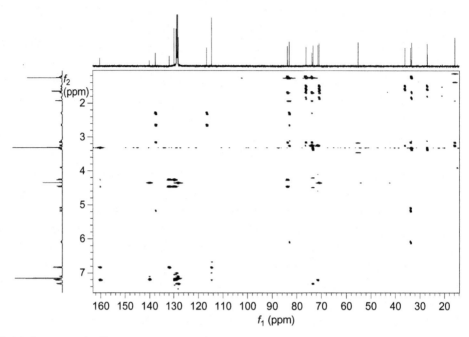

FIG. 5.2.2.6   The 2-D ^{1}H-^{13}C gHMBC NMR spectrum of molecule 5.2.2 in benzene-d_6.

TABLE 5.2.2.1   Predicted Multiplicities for the ^{1}H's of Molecule 5.2.2

Site in Molecule	Expected Multiplicity
2	d^2
3 (OH)	s
4	d^2
5	q, dt
6	d^4
7	$2 \times d^3$
8	d^4
9	$2 \times d^2$
10	s
11	$2 \times d$
13/17	d^a
14/16	d^a
18	s
19	$2 \times d^4$
20	$2 \times d^5$

TABLE 5.2.2.1 Predicted Multiplicities for the ^{1}H's of Molecule 5.2.2—cont'd

Site in Molecule	Expected Multiplicity
21	$2 \times d^3$
22	$2 \times d$
24/28	d^a
25/27	t^a
26	t

^{a}Double intensity.

TABLE 5.2.2.2 Numbered Sites of Molecule 5.2.2

Type of Site	Site Number
CH_3 (methyl)	10, 18
CH_2 (methylene)	5, 7, 9, 11, 19, 20, 21, 22
CH (methine)	2, 4, 6, 8, 13/17, 14/16, 24/28, 25/27, 26
C_{np} (nonprotonated)	3, 12, 15, 23
Numbered heteronucleus	1 (O)
Protonated heteronucleus	3OH

TABLE 5.2.2.3 ^{1}H and ^{13}C NMR Signals of Molecule 5.2.1 Listed by Group Type

^{1}H Signal (ppm)	^{13}C Signal (ppm)	Group Type
–	160.1	Nonprotonated
–	139.9	Nonprotonated
6.08	137.4	Methine
–	131.8	Nonprotonated
7.21	129.7^a	Methine
7.17	$128.9^{a,b}$	Methine
7.33	128.1^a	Methine
7.10	128.0	Methine
5.09 & 5.17	116.5	Methylene
6.83	114.6^a	Methine
3.24	83.7	Methine
3.16	82.9	Methine
3.12	76.2	Methine
–	73.7	Nonprotonated

Continued

TABLE 5.2.2.3 ^{1}H and ^{13}C NMR Signals of Molecule 5.2.1 Listed by Group Type—cont'd

^{1}H Signal (ppm)	^{13}C Signal (ppm)	Group Type
4.33 (2)	73.3	Methylene
4.24 & 4.44	71.3	Methylene
3.31 & 3.37	70.7	Methylene
3.32	55.2	Methyl
1.21 & 1.69	36.0	Methylene
2.30 & 2.64	33.7	Methylene
1.49 & 1.57	33.3	Methylene
1.67 & 1.84	26.9	Methylene
1.23	15.6	Methyl

^{a}*Doubly intense.*
^{b}*Extra ^{13}C signals observed near this signal.*

We begin (again, as with molecule 5.2.1) with the allyl group. The most downfield methylene set of two ^{1}H signals correlating with one ^{13}C signal is midfield in the ^{1}H chemical shift range and downfield in the ^{13}C chemical shift range. We assign the midfield ^{1}H signals at 5.09 & 5.17 ppm that correlate with the ^{13}C signal at 116.5 ppm to site 9: $\delta_H = 5.09$ & 5.17 ppm and $\delta_C = 116.5$ ppm.

The H9 signals at 5.09 & 5.17 ppm share strong COSY interactions with the signal of a downfield methine ^{1}H signal at 6.08 ppm. We assign the 6.08 ppm ^{1}H signal to site 8: $\delta_H = 6.08$ ppm and $\delta_C = 137.4$ ppm.

The H8 signal at 6.08 ppm shares a COSY cross peak with a methylene ^{1}H signal at 2.64 ppm. We assign this methylene ^{1}H signal to site 7: $\delta_H = 2.30$ & 2.64 ppm and $\delta_C = 33.7$ ppm.

Both the H7 signals at 2.30 & 2.64 ppm correlate in the COSY spectrum with the midfield methine ^{1}H signal at 3.16 ppm. We assign this methine ^{1}H signal to site 2: $\delta_H = 3.16$ ppm and $\delta_C = 82.9$ ppm.

We assign the more upfield of the two methyl ^{1}H signals at 1.23 ppm to site 10: $\delta_H = 1.23$ ppm and $\delta_C = 15.6$ ppm.

The H10 signal at 1.23 ppm and the H2 signal at 3.24 ppm both share gHMBC cross peaks with the only nonprotonated midfield ^{13}C signal at 73.7 ppm. We assign this signal to site 3: $\delta_C = 73.7$ ppm. The cross peak shared by the signals of H10 and C3 can be differentiated from other nearby gHMBC cross peaks by noting how skinny it is in the vertical dimension (f_2 axis). The ^{13}C shift of this very elongated gHMBC cross peak can be interpolated by averaging the extremes along the ^{13}C chemical shift axis (f_1). Complicating this methodology is the presence of the nearby gHMBC cross peak between the as-yet unassigned methylene ^{1}H signal at 1.21 ppm which correlates with the as-yet unassigned methine ^{13}C signal at 76.2 ppm.

The ^{1}H signal at 1.92 ppm shares a gHMBC cross peak with the C3 signal at 73.7 ppm. We assign this ^{1}H signal to the hydroxyl group on site 3 (OH): $\delta_H = 1.92$ ppm.

The site 3 hydroxyl signal is found to correlate with other ^{13}C signals in the gHMBC spectrum. In particular, the site 3 hydroxyl ^{1}H signal at 1.92 ppm shares gHMBC cross peaks with the C2 signal at 82.9 ppm and also with a second midfield methine ^{13}C signal at 83.7 ppm. We assign the midfield methine ^{13}C signal to site 4: $\delta_H = 3.24$ ppm and $\delta_C = 83.7$ ppm.

We observe a strong gHMBC correlation between methylene ^1H signals at 1.21 & 1.69 ppm and the C4 signal at 83.7 ppm. We assign the methylene ^1H signals to site 5: $\delta_H = 1.21$ & 1.69 ppm and $\delta_C = 36.0$ ppm. The availability of a sole low energy conformation for the site 1–6 ring causes the ring to adopt a single, static chair conformation, narrowing the C5 resonance and hence generating more observed cross peaks in the gHMBC spectrum. To emphasize this important point: Molecule 5.2.2 generates a gHMBC spectrum with more cross peaks involving the site 1–6 ring because (1) molecular conformation, include bond and dihedral angles, varies less in molecule 5.2.2 than in molecule 5.2.1, and (2) the ^{13}C chemical resonances are sharper, thereby confining all the volume of the gHMBC cross peak to a smaller region, which for a given volume makes the cross peak taller—the very thing to which plotting of contours above a certain minimum threshold is sensitive.

The H5 signals at 1.21 & 1.69 ppm correlate with an unassigned midfield methine ^{13}C signal at 76.2 ppm. We assign this methine ^{13}C signal to site 6: $\delta_H = 3.12$ ppm and $\delta_C = 76.2$ ppm.

The midfield methyl $^1H/^{13}C$ signal pair with a ^1H chemical shift of 3.32 ppm and at a ^{13}C chemical shift of 55.2 ppm is assigned to site 18: $\delta_H = 3.32$ ppm and $\delta_C = 55.2$ ppm.

The H18 signal at 3.32 ppm correlates with the most downfield ^{13}C signal in the gHMBC spectrum at 160.1 ppm. This ^{13}C signal is assigned to site 15: $\delta_C = 160.1$ ppm.

The H13/17 signals are expected to correlate strongly with the C15 signal in the gHMBC spectrum while the H14/16 signals should only do so weakly. The lone pair of the site 15 oxygen atom donates electron density to sites 14/16, thereby shifting the site 14/16 signals upfield from those of their site 13/17 counterparts. The aromatic ^1H signals at 7.21 ppm correlate strongly with the C15 signal at 160.1 ppm, while the more upfield aromatic ^1H signal at 6.83 ppm correlates only moderately with the C15 signal. We assign the more strongly correlated and downfield aromatic ^1H signal to sites 13/17: $\delta_H = 7.21$ ppm and $\delta_C = 129.7$ ppm. We assign the upfield aromatic ^1H signal to sites 14/16: $\delta_H = 6.83$ ppm and $\delta_C = 114.6$ ppm.

The H14/16 signals are expected to correlate strongly with the C12 signal in the gHMBC spectrum. We observe a strong gHMBC cross peak between the H14/16 signal at 6.83 ppm and a nonprotonated ^{13}C signal at 131.8 ppm. We assign this nonprotonated ^{13}C signal to site 12: $\delta_C = 131.8$ ppm.

A pair of midfield methylene ^1H signals at 4.24 & 4.44 ppm is observed to share gHMBC cross peaks with the C12 signal at 131.8 ppm. We assign these methylene ^1H signals to site 11: $\delta_H = 4.24$ & 4.44 ppm and $\delta_C = 71.3$ ppm. We confirm our site 11 assignment by noting the presence of a cross peak shared by the H11 signals at 4.24 & 4.44 ppm and the C4 signal at 83.7 ppm in the gHMBC spectrum.

The site 26 aromatic ^1H signal can be identified in the 1-D ^1H NMR spectrum of molecule 5.2.2 because it is the only aromatic ^1H signal that has half of the intensity of all of the other aromatic ^1H signals of the solute (the ^1H signal of benzene-d_5,h_1 is found at 7.16 ppm, complicating the integrals). We write for site 26: $\delta_H = 7.10$ ppm and $\delta_C = 128.0$ ppm.

Of the two unassigned aromatic ^1H signals at 7.17 and 7.33 ppm, the H26 signal at 7.10 ppm shares a strong COSY correlation with the more upfield (7.17 ppm) ^1H signal of the two. We assign this more upfield aromatic ^1H signal to sites 25/27: $\delta_H = 7.17$ ppm and $\delta_C = 128.9$ ppm. We assign the last unassigned aromatic $^1H/^{13}C$ signal pair to sites 24/28: $\delta_H = 7.33$ ppm and $\delta_C = 128.1$ ppm.

The H25/27 signal at 7.17 ppm correlates in the gHMBC spectrum with the last unassigned aromatic ^{13}C signal at 139.9 ppm. We assign this ^{13}C signal to site 23: $\delta_C = 139.9$ ppm.

The H24/28 signal at 7.33 ppm shares a gHMBC cross peak with a midfield methylene ^{13}C signal at 73.3 ppm. We assign this methylene ^{13}C signal to site 22: $\delta_H = 4.33$ ppm (both ^{1}H's) and $\delta_C = 73.3$ ppm. We confirm this assignment by noting the presence of a gHMBC cross peak between the H22 signal at 4.33 ppm and the C23 signal at 139.9 ppm.

The H22 signal at 4.33 ppm correlates in the gHMBC spectrum with another midfield methylene ^{13}C signal at 70.7 ppm. We assign this midfield ^{13}C signal to site 21: $\delta_H = 3.31$ & 3.37 ppm and $\delta_C = 70.7$ ppm. The newly assigned H21 signals at 3.31 & 3.37 ppm are observed to share gHMBC correlations with the C22 signal at 73.3 ppm, thus confirming our assignment of the NMR signals associated with site 21.

The H21 signals at 3.31 & 3.37 ppm correlate in the gHMBC spectrum with the last two unassigned methylene ^{13}C signals at 26.9 and 33.3 ppm, providing no clear differentiation between sites 19 and 20. However, the H21 signals *are* observed to correlate in the COSY spectrum with a pair of methylene ^{1}H signals at 1.67 & 1.84 ppm. We assign these new methylene ^{1}H signals to site 20: $\delta_H = 1.67$ & 1.84 ppm and $\delta_C = 26.9$ ppm.

The NMR signals of the last methylene group are assigned to site 19: $\delta_H = 1.49$ & 1.57 ppm and $\delta_C = 33.3$ ppm. We note that the H6 signal at 3.12 ppm shares a weak gHMBC correlation with the C19 signal at 33.3 ppm. The more upfield H5 signal (at 1.21 ppm) correlates strongly in the gHMBC spectrum with the C19 signal at 33.3 ppm. The H19 signals at 1.49 & 1.57 ppm also share gHMBC cross peaks with the C6 signal at 76.2 ppm, with the C21 signal at 70.7 ppm, with the C5 signal at 36.0 ppm, and with the C20 signal at 26.9 ppm.

5.2.3 Discussion of the Epimeric Consequences of Site 2 Inversion Between Molecules 5.2.1 and 5.2.2

For molecule 5.2.2, the right-handedness of the site 2 stereocenter allows the sites 1–6 ring to adopt a low energy chair conformation in which H2, H4, and H6 are axial, below the plane of the ring as shown in Fig. 5.2.2.1. The site 10 methyl group is also axial but above the plane of the ring as shown in Fig. 5.2.2.1. We expect the gHMBC cross peak between the signals of H2 and C10 to be stronger for molecule 5.2.2 than for molecule 5.2.1. For molecule 5.2.1, the gHMBC cross peak between the signals of H2 and C10 appears in Fig. 5.2.1.6 at the ^{1}H chemical shift of 3.79 ppm and a ^{13}C chemical shift of 22.4 ppm. For molecule 5.2.2, the same cross peak is found at a ^{1}H shift of 3.16 ppm and the ^{13}C shift of 15.6 ppm. This gHMBC cross peak for molecule 5.2.1 is observed to be less intense than for molecule 5.2.2. This is consistent with our expectation that the dominant chair conformation of molecule 5.2.2 promotes orbital overlap that results in a strong coupling between H2 and C10.

Assuming that the site 1–6 ring is in a chair conformation for molecule 5.2.1, we note that the H2 signal appears at a more downfield chemical shift when it occupies an equatorial position compared with the shift we observe for H2 when it occupies an axial position in molecule 5.2.2. This chemical shift difference for the H2 signals is consistent with our expectation that an equatorial ^{1}H signal will be found at downfield (at a greater ppm value) than its equatorial counterpart.

5.3 REBAUDIOSIDE-A IN D$_2$O

Rebaudioside-A is a naturally occurring glycoside of steviol, meaning that it is the molecule steviol with attached sugars. The structure of rebaudioside-A is shown in Fig. 5.3.1 and the structure of steviol is shown in Fig. 5.3.2. Rebaudioside-A has eight rings, two double bonds, one of which is part of an ester group, 14 hydroxyl groups, seven ether linkages, and two methyl groups. With an index of hydrogen deficiency of ten, the empirical formula of Rebaudioside-A is $C_{44}H_{70}O_{23}$. Counting atoms is an important part of the accounting we perform prior to examining any data. The steviol residue (so named because it has been linked to sugars via the formation of an ether and an ester linkage with the elimination of water, hence the residue denotation) contributes $C_{20}H_{28}O_3$ (including the ether and ester oxygens in the steviol portion). Each unfunctionalized sugar ring contributes $C_6H_{11}O_5$, but ring II is functionalized at sites 2″ and 3″ with two ether linkages, so ring II only contributes nine hydrogens instead of eleven.

The NMR signals of the molecule are largely divided between the mostly upfield signals of the steviol residue (however, C16 and C17 are sp^2-hybridized) and the midfield signals from four cyclic sugars. Two factors make this problem challenging: (1) overlap of the NMR signals from the four sugar rings is extensive, and (2) the central steviol portion of the molecule exhibits apparent flexibility near sites 1, 2, and 3, near sites 6 and 7, and also near sites 11 and 12. Do not suppose that the flexibility is intuitively obvious: the apparent molecular flexibility is only discovered while making assignments. Molecular rearrangements on the NMR time scale prevent the observation of interactions involving those J-couplings that vary as the result of the molecule adopting different conformations, thereby producing few, if any, observable cross peaks involving ^3J's for fluxional sites. Signals from all of the hydroxyl ^{1}H's

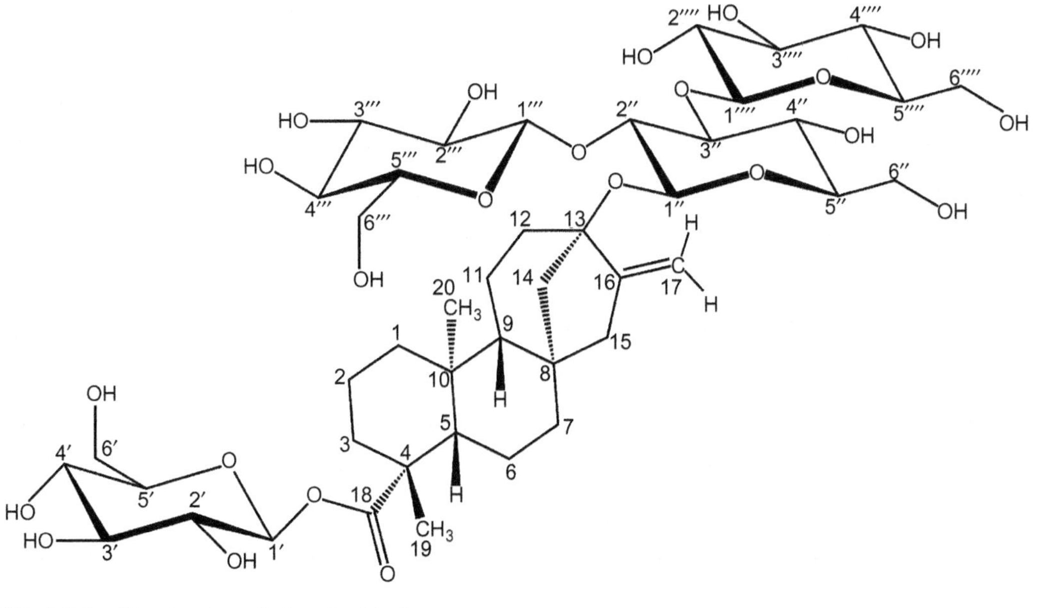

FIG. 5.3.1 The structure of rebaudioside-A.

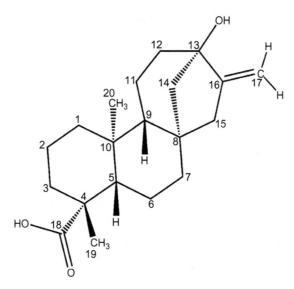

FIG. 5.3.2   The structure of steviol, from whence rebaudioside-A is derived.

are not observed owing to solvent exchange, reducing, perhaps gratifyingly, the number of ^{1}H's whose signals we must assign from 70 to a paltry 56. Information furnished by the 2-D ^{1}H-^{1}H TOCSY experiment proves instrumental in assigning the signals from the sugar rings.

This sample was provided to the author by Dr. Prashant Chopade in 2012, who was at the time working for Professor Alan Myerson in the Department of the Chemical Engineering at the Massachusetts Institute of Technology (Chopade et al., 2016). The sample was dissolved in deuterated water and this one sample was used to generate all of the data from which the figures in this section are derived. Fig. 5.3.3 shows the 1-D ^{1}H NMR spectrum of the sample. Fig. 5.3.4 shows the 1-D ^{13}C NMR spectrum of Rebaudioside-A. Notice that the peak picking algorithm has selected a spurious ^{13}C signal at 100.5 ppm whose amplitude in the power spectrum (not shown) is above the peak picking threshold. We can easily identify this signal as an artifact because its phase differs from the rest of our observed 1-D ^{13}C NMR signals. Also notice that the signals from methylene and methine carbon sites are broadened enough to appear shorter than the signals arising from their nonprotonated counterparts in the 1-D ^{13}C spectrum. The large mass of the Rebaudioside A molecule causes the molecule to tumble slowly enough in solution that attached ^{1}H's provide efficient T_2 relaxation, resulting in broader ^{13}C resonances (if just slow tumbling alone were responsible for broadening of the ^{13}C signals, then nonprotonated ^{13}C signals would be similarly affected). The 2-D ^{1}H-^{1}H COSY spectrum is shown in Fig. 5.3.5. Three 2-D ^{1}H-^{1}H TOCSY spectra are shown in Figs. 5.3.6–5.3.8 with mixing times of 40, 80, and 120 ms. The 2-D ^{1}H-^{13}C HSQC NMR spectrum of the sample is shown in Fig. 5.3.9. Three expanded portions of the HSQC spectrum appear in Figs. 5.3.10–5.3.12. The 2-D ^{1}H-^{13}C gHMBC NMR spectrum is found in Fig. 5.3.13. An expanded portion of the gHMBC spectrum appears in Fig. 5.3.14.

As is our custom, we predict the expected ^{1}H NMR signal multiplicities and place these predictions into Table 5.3.1. We note that as molecular complexity increases, so does resonance overlap, and therefore our multiplicity predictions are of a correspondingly diminished utility. We parse through the molecular structure, recording the various types of carbon atoms

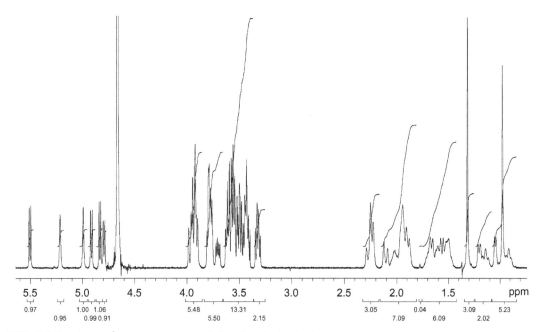

FIG. 5.3.3 The 1-D ¹H NMR spectrum of rebaudioside-A in D₂O.

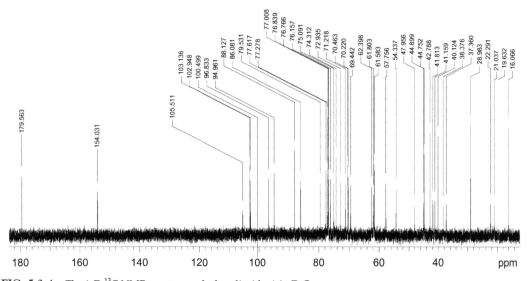

FIG. 5.3.4 The 1-D ¹³C NMR spectrum of rebaudioside-A in D₂O.

in the molecule, placing our results in Table 5.3.2. Finally, we measure the NMR signals we observe in the 1-D spectra as well as in the HSQC spectrum and place our distillation of these measurements into Table 5.3.3. We have added an additional column to Table 5.3.3 compared with other tables containing chemical shifts to allow us to discuss unassigned NMR signals with letter labels instead of possible molecular site numbers. The letter labels will help

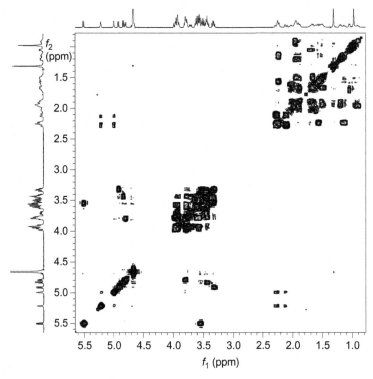

FIG. 5.3.5   The 2-D ^{1}H-^{1}H COSY NMR spectrum of rebaudioside-A in D_2O.

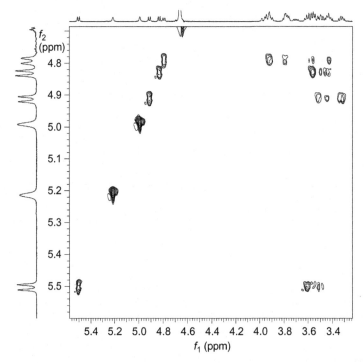

FIG. 5.3.6   The 2-D ^{1}H-^{1}H TOCSY NMR spectrum of rebaudioside-A in D_2O, obtained with a mixing time of 40 ms.

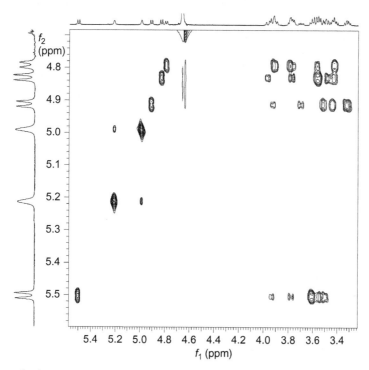

FIG. 5.3.7 The 2-D ^{1}H-^{1}H TOCSY NMR spectrum of rebaudioside-A in D₂O, obtained with a mixing time of 80 ms.

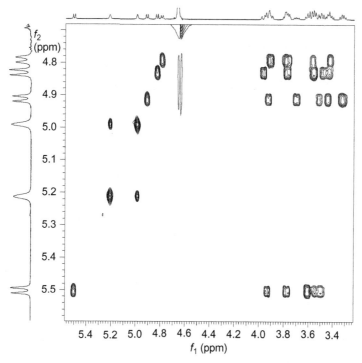

FIG. 5.3.8 The 2-D ^{1}H-^{1}H TOCSY NMR spectrum of rebaudioside-A in D₂O, obtained with a mixing time of 120 ms.

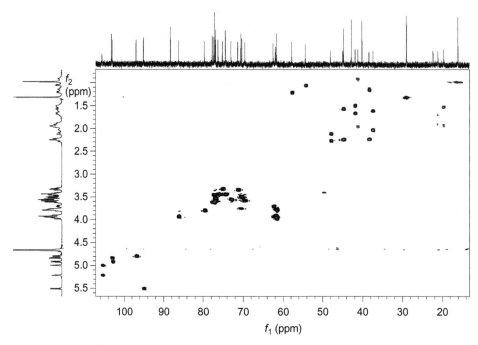

FIG. 5.3.9 The 2-D ^{1}H-^{13}C HSQC NMR spectrum of rebaudioside-A in D$_2$O.

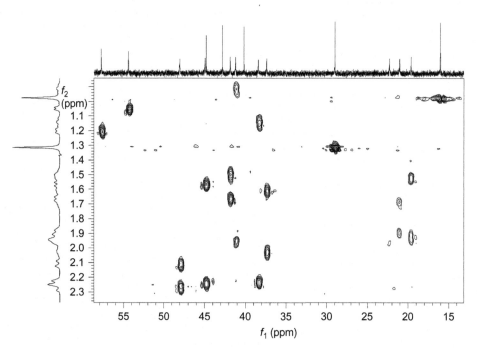

FIG. 5.3.10 An expanded portion of the 2-D ^{1}H-^{13}C HSQC NMR spectrum of rebaudioside-A in D$_2$O, showing the upfield chemical shift region.

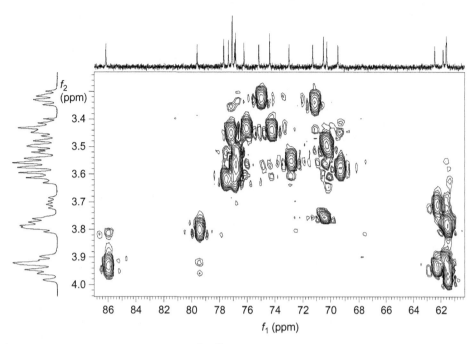

FIG. 5.3.11 An expanded portion of the 2-D ¹H-¹³C HSQC NMR spectrum of rebaudioside-A in D₂O, showing the midfield chemical shift region.

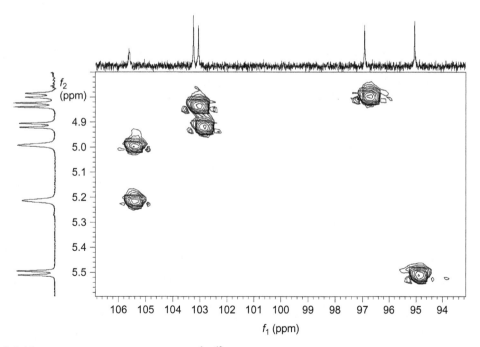

FIG. 5.3.12 An expanded portion of the 2-D ¹H-¹³C HSQC NMR spectrum of rebaudioside-A in D₂O, showing the downfield chemical shift region.

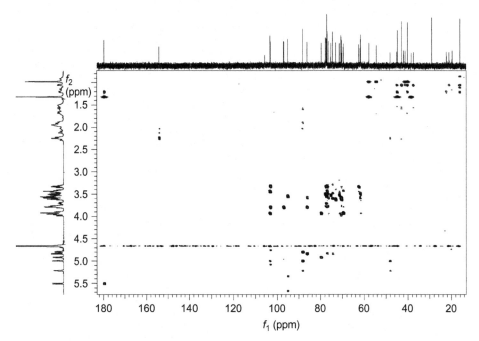

FIG. 5.3.13   The 2-D ^{1}H-^{13}C gHMBC NMR spectrum of rebaudioside-A in D$_2$O.

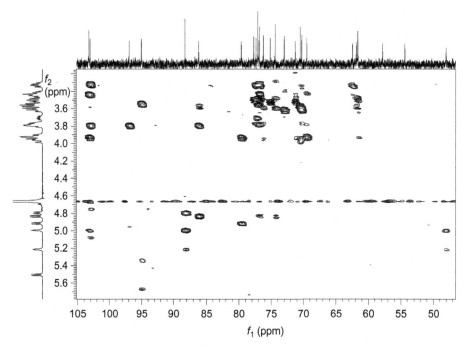

FIG. 5.3.14   An expanded portion of the 2-D ^{1}H-^{13}C gHMBC NMR spectrum of rebaudioside-A in D$_2$O.

TABLE 5.3.1 Predicted Multiplicities for the ^{1}H's of Rebaudioside-A

Site in Molecule	Expected Multiplicity
1	$2 \times d^3$
2	$2 \times d^5$
3	$2 \times d^3$
5	d^2
6	$2 \times d^4$
7	$2 \times d^3$
9	d^2
11	$2 \times d^4$
12	$2 \times d^3$
14	$2 \times d$
15	$2 \times d$
17	$2 \times d^a$
19	s
20	s
1′	d
2′	d^2
3′	d^2
4′	d^2
5′	d^3
6′	$2 \times d^2$
1″	d
2″	d^2
3″	d^2
4″	d^2
5″	d^3
6″	$2 \times d^2$
1‴	d
2‴	d^2
3‴	d^2
4‴	d^2
5‴	d^3

Continued

TABLE 5.3.1   Predicted Multiplicities for the ^{1}H's of Rebaudioside-A—cont'd

Site in Molecule	Expected Multiplicity
6$'''$	$2\times d^2$
1$''''$	d
2$''''$	d^2
3$''''$	d^2
4$''''$	d^2
5$''''$	d^3
6$''''$	$2\times d^2$

^{a} $^2J_{HH}$ expected to be small.

TABLE 5.3.2   Numbered Sites of Rebaudioside-A

Type of Site	Site Number
CH$_3$ (methyl)	19, 20
CH$_2$ (methylene)	1, 2, 3, 6, 7, 11, 12, 14, 15, 17, 6$'$, 6$''$, 6$'''$, 6$''''$
CH (methine)	5, 9, 1$'$, 2$'$, 3$'$, 4$'$, 5$'$, 1$''$, 2$''$, 3$''$, 4$''$, 5$''$, 1$'''$, 2$'''$, 3$'''$, 4$'''$, 5$'''$, 1$''''$, 2$''''$, 3$''''$, 4$''''$, 5$''''$
C$_{np}$ (nonprotonated)	4, 8, 10, 13, 16, 18

TABLE 5.3.3 ^{1}H and ^{13}C NMR Signals of Rebaudioside-A Listed by Carbon Type

Label	^{1}H Signal (ppm)	^{13}C Signal (ppm)	Group Type
	–	179.6	Nonprotonated
	–	154.0	Nonprotonated
	4.99 & 5.22	105.5	Methylene
	4.84	103.1	Methine
	4.92	102.9	Methine
	4.79	96.8	Methine
	5.50	95.0	Methine
	–	88.1	Nonprotonated
D	3.93	86.1	Methine
E	3.80	79.5	Methine
F	3.62	77.6	Methine
G	3.45	77.3	Methine
H,I	3.56, 3.63	77.0^a	Methine

TABLE 5.3.3 ^{1}H and ^{13}C NMR Signals of Rebaudioside-A Listed by Carbon Type—cont'd

Label	^{1}H Signal (ppm)	^{13}C Signal (ppm)	Group Type
J	3.51	76.84	Methine
K	3.58	76.77	Methine
L	3.42	76.2	Methine
M	3.32	75.1	Methine
N	3.44	74.3	Methine
O	3.55	72.9	Methine
P	3.33	71.2	Methine
Q	3.75	70.5	Methine
R	3.48	70.2	Methine
S	3.58	69.4	Methine
T	3.71 & 3.925	62.4	Methylene
U	3.78 & 3.92	61.8	Methylene
V,W	3.775 & 3.94, 3.77 & 3.96	61.6^a	Methylene
	1.21	57.8	Methine
	1.06	54.3	Methine
	2.12 & 2.27	48.0	Methylene
A	1.57 & 2.25	44.9	Methylene
	–	44.8	Nonprotonated
	–	42.8	Nonprotonated
B	1.51 & 1.66	41.8	Methylene
	0.92 & 1.96	41.2	Methylene
	–	40.1	Nonprotonated
	1.15 & 2.24	38.4	Methylene
C	1.60 & 2.04	37.4	Methylene
	1.32	29.0	Methyl
	1.93 & 1.97^b	22.3^c	Methylene
	1.68 & 1.90	21.0	Methylene
	1.52 & 1.94	19.6	Methylene
	0.98	16.1	Methyl

^{a}Doubly intense.
^{b}Determined via the integral of the 1-D ^{1}H spectrum.
^{c}Assumed to be fluxional methylene ^{13}C signal instead of the more downfield C$_{np}$ signal candidates at 40.1, 42.8, and 44.8 ppm.

facilitate our discussion of the signals as we pair signals with numbered sites. Use of letters to label unassigned resonances helps avoid confusion between numbered sites and resonances that have been numbered in order of their appearance in the frequency spectrum.

We are only able to observe 12 pairs of cross peaks from the 13 methylene groups in Rebaudioside-A. We are able to determine the approximate chemical shifts of the missing methylene ^{1}H signals thanks to quantitative integral values in the 1-D ^{1}H NMR spectrum by noting that the seven ^{1}H signals between 1.85 and 2.15 ppm generate only five cross peaks in the HSQC spectrum. We are able to infer the chemical shifts of the missing methylene ^{1}H signals by determining where the HSQC-silent ^{1}H signals reside in the 1-D spectrum. We assign this pair of ^{1}H signals to the ^{13}C resonance at 22.3 ppm, which is too far upfield for a nonprotonated ^{13}C in the absence of profound ring strain or proximity to the middle of an aromatic ring.

We begin by pairing sites and signals found in the upfield region of the HSQC spectrum. In this region we observe two methine ^{1}H/^{13}C signal pairs, two methyl signal pairs, and 16 cross peaks from eight (with a ninth inferred) methylene groups. Assigning sites to signals for the methyls and methines will likely be easier than for the nine upfield methylenes.

Our entry into the steviol portion of the molecule is afforded by the ester group's ^{13}C signal. Site 18 of our molecule has a single carbonyl (in the ester functional group) that joins the lower left portion of the steviol residue to sugar ring I (singly primed sites). We assign the most downfield ^{13}C signal at 179.6 ppm to site 18: $\delta_C = 179.6$ ppm.

The methyl ^{1}H signal at 1.32 ppm is observed to correlate in the gHMBC spectrum with the C18 signal at 179.6 ppm. We assign this methyl ^{1}H signal to site 19: $\delta_H = 1.32$ ppm and $\delta_C = 29.0$ ppm. The proximity of the oxygens from the nearby site 18 ester group is consistent with the assignment of the more downfield of the two methyl NMR signals to site 19.

The methine ^{1}H signal at 1.21 ppm shares a gHMBC cross peak with the C18 signal at 179.6 ppm. We assign this methine ^{1}H signal to site 5: $\delta_H = 1.21$ ppm and $\delta_C = 57.8$ ppm. If we assume the ring composed of sites 1, 2, 3, 4, 5, and 10 has adopted a chair conformation, we expect a strong coupling from H5 to C18 because the coupling will be a trans-$^3J_{CH}$. Even if the methylene group at site 2 flops over and forms a skew boat, conformational variation affecting the H5-C18 dihedral angle is expected to be minimal due to the rigidity of the two rings that share sites 5 and 10. Signal dispersion is sufficient near the H5 signal to allow us to be confident in this assignment.

The site 19 methyl ^{1}H signal at 1.32 ppm and the site 5 methine ^{1}H signal at 1.21 ppm both correlate with a nonprotonated ^{13}C signal at 44.8 ppm. Although a methylene ^{13}C signal is nearby at 44.9 ppm, we hold that the signals of the H19's and that of H5 will both correlate with the nonprotonated C4 signal, because C4 is only two bonds distant from H5 and the H19's. The ^{13}C signal at 44.8 ppm in the gHMBC spectrum is the only nonprotonated ^{13}C signal that correlates in the gHMBC with the signals of both H5 and the H19's and so we write for site 4: $\delta_C = 44.8$ ppm.

The H19 signal at 1.32 ppm shares a gHMBC cross peak with the ^{13}C signal from an upfield methylene group at 38.4 ppm. We assign this methylene ^{13}C signal to site 3: $\delta_H = 1.15$ & 2.24 ppm and $\delta_C = 38.4$ ppm.

Because we have identified the NMR signals of one of the two methyl groups (site 19), the other methyl NMR signals in the last row of Table 5.3.3 are assigned to site 20: $\delta_H = 0.98$ ppm and $\delta_C = 16.1$ ppm.

We also use the process of elimination to assign the other upfield methine ^{1}H/^{13}C signal pair to site 9: $\delta_H = 1.06$ ppm and $\delta_C = 54.3$ ppm since none of the methine groups in the sugars will generate ^{1}H signals with such a low chemical shift. We observe a gHMBC cross peak between the H9 signal at 1.06 ppm and the C20 signal at 16.1 ppm, consistent with our assignments thus far.

In the gHMBC spectrum, the H20 signal at 0.98 ppm correlates with the C5 signal at 57.8 ppm, with the C9 signal at 54.3 ppm, with a methylene ^{13}C signal at 41.8 ppm, and with a nonprotonated ^{13}C signal at 40.1 ppm. We assign the methylene ^{13}C signal at 41.8 ppm to site 1: $\delta_H = 1.50$ & 1.66 ppm and $\delta_C = 41.8$ ppm. We assign the nonprotonated ^{13}C signal at 40.1 ppm to site 10: $\delta_C = 40.1$ ppm.

If we examine the nine methylene groups in the steviol portion of our molecule, we can predict that the NMR signals of sites 2, 6, and 11 will be most upfield, insofar as each methylene is bracketed by methylenes or methines instead of other more electron-density-hungry groups. We expect that the last three methylene rows in Table 5.3.3 will be linked with sites 2, 6, and 11.

The H9 signal at 1.06 ppm is observed to share a gHMBC correlation with the upfield methylene signal at 21.0 ppm. We assign this methylene ^{13}C signal to site 11: $\delta_H = 1.68$ & 1.90 and $\delta_C = 21.0$ ppm.

The H5 signal at 1.21 ppm also correlates with one of the most upfield methylene ^{13}C resonances at a chemical shift of 22.3 ppm. We assign this ^{13}C signal to site 6: $\delta_H = 1.93$ & 1.97 ppm and $\delta_C = 22.3$ ppm.

By the process of elimination we assign the last of the most upfield NMR signals to site 2: $\delta_H = 1.52$ & 1.94 ppm and $\delta_C = 19.6$ ppm. While we may be somewhat dismayed to note that our newly assigned C2 signal has a chemical shift involved in no gHMBC cross peaks, we can take some solace from the fact that (1) site 2 likely flops back and forth as its ring converts its conformation between a chair and a skew boat, and (2) the H1's and H3's are both coupled to and their signals split by three spins and so are expected to generate signals that are more spread out, lowering the heights of their C2-signal-involved gHMBC cross peaks below the plotting threshold.

The H9 signal at 1.06 ppm is observed to share a gHMBC cross peak with the last unassigned upfield nonprotonated ^{13}C resonance at 42.8 ppm. This ^{13}C resonance is assigned to site 8: $\delta_C = 42.8$ ppm. The only other nonprotonated ^{13}C signal remaining is for site 13, which is alpha to oxygen and expected to have a chemical shift greater than 60 ppm.

Having finally assigned our way around the molecule until we are close to the carbon-carbon double bond between C16 and C17, we take this opportunity record an assignment we could have easily used as an entry point. We assign the most downfield methylene NMR signals to site 17: $\delta_H = 4.99$ & 5.22 ppm and $\delta_C = 105.5$ ppm.

The second most downfield ^{13}C resonance is assigned, on no other basis than its downfield chemical shift, to site 16: $\delta_C = 154.0$ ppm. We should not be too disappointed not to see a gHMBC correlation between the signals of the H17's and that of C16 because C17 is sp^2-hybridized and so the $^2J_{CH}$ between H17 and C16 is expected to be small.

The H17 signals do, however, correlate in the gHMBC spectrum with a methylene ^{13}C signal at 48.0 ppm. This methylene ^{13}C signal is assigned to site 15: $\delta_H = 2.12$ & 2.27 ppm and $\delta_C = 48.0$ ppm.

The H17 signals at 4.99 & 5.22 ppm also share gHMBC cross peaks with a midfield nonprotonated ^{13}C signal at 88.1 ppm. This ^{13}C signal is assigned to site 13: $\delta_C = 88.1$ ppm. This was the last unassigned nonprotonated ^{13}C signal, and because it is associated with a ^{13}C bonded to oxygen, the chemical shift is noted as being appropriately greater than 60 ppm.

At this point, the steviol residue has only three methylene assignments yet to be made, namely, those for sites 7, 12, and 14. We have added the labels A, B, and C to the rows in

Table 5.3.3 containing the three steviol methylene NMR signals that remain unassigned. Unfortunately, there is a dearth of gHMBC cross peaks associated with the NMR signals from sites 7, 12, and 14. Also, resonance overlap along the ^{1}H chemical shift axis renders the COSY spectrum challenging to interpret unambiguously. The most downfield signal from the remaining three methylene pairs has ^{1}H signals at 1.57 & 2.25 ppm. The downfield ^{1}H signal from this pair at 2.25 ppm correlates in the gHMBC spectrum with the C16 signal at 154.0 ppm and with the C15 signal at 48.0 ppm. Note that no other already-assigned ^{1}H signal near 2.25 ppm (H3, H15) is a viable candidate for generating a gHMBC cross peak with the C15 signal. There is a significant intensity difference between the two gHMBC cross peaks that appear to correlate the H15 signals at 2.12 & 2.27 ppm to the C16 signal at 154.0 ppm. It is unlikely that the upfield H15 signal at 2.12 ppm correlates much more weakly with the C16 signal than its downfield counterpart at 2.27 ppm. The gHMBC cross peaks between the two H15 signals and that of C16 are likely comparable in intensity. Therefore the strong intensity difference between the gHMBC cross peak shared by the C16 signal and the ^{1}H signal(s) at 2.25/7 ppm versus the upfield H15 signal at 2.12 ppm is ascribed to a cross peak between the C16 signal and the as-of-yet unassigned methylene ^{1}H signal at 2.25 ppm (label A). The upfield ^{1}H signal at 1.57 ppm from this unassigned methylene group is observed to share gHMBC cross peaks with the C13 signal at 88.1 ppm and with the C8 signal at 42.8 ppm. The methylene group most likely generating the ^{1}H signals participating in the gHMBC cross peaks described previously in this paragraph is the site 14 methylene group, because the H14's are two bonds from C8 and C13 and also three bonds from C15 and C16. In comparison we must consider the bond proximity of the H7's and the H12's to C8, C13, C15, and C16. The H7's are two bonds from C8, three bonds from C15, four bonds from C16, and five bonds from C13. The H12's are two bonds from C13, three bonds from C16, and four bonds from C8 and C15. We assign the label A NMR signals to site 14: $\delta_H = 1.57$ & 2.25 ppm and $\delta_C = 44.9$ ppm.

The label C methylene ^{1}H signal at 2.04 ppm correlates in the gHMBC spectrum with the C16 signal at 154.0 ppm and with the C13 signal at 88.1 ppm. This label C methylene ^{1}H signal is assigned to site 12: $\delta_H = 1.60$ & 2.04 ppm and $\delta_C = 37.4$ ppm. Consistent with the assignment of site 12 to the label C NMR signals, the H9 signal at 1.06 ppm shares a gHMBC cross peak with the C12 signal at 37.4 ppm. We have unfortunately confirmed little in noting the presence of the gHMBC cross peak between the H9 signal at 1.06 ppm and the putative C12 at 41.2 ppm, for we must admit that H9 can reasonably be expected to couple to C7, C12, and C14 more or less equally since it is three bonds from each of the carbon atoms C7, C12, and C14.

The last set of the steviol residue's methylene NMR signals is assigned to site 7: $\delta_H = 0.92$ & 1.96 ppm and $\delta_C = 41.2$ ppm. The lack of gHMBC cross peaks that might be used to confirm our last few assignments is attributed to the combination of molecular flexibility and resonance broadening through coupling to other nearby spins.

We now move on to the sugar rings. We can begin by examining the midfield portion of the HSQC spectrum to familiarize ourselves with the challenge ahead. We are able to identify the ^{1}H/^{13}C cross peaks from spins of the anomeric sites by looking for ^{13}C signals with chemical shifts near 100 ppm. Excluding the already-assigned site 17 signals from the terminal vinyl group, we can readily identify the NMR signals of the four anomeric ^{1}H/^{13}C spin pairs with ^{1}H chemical shifts of 4.79, 4.83, 4.92, and 5.50 ppm, and ^{13}C chemical shifts of 96.8, 103.1, 102.9, and 95.0 ppm, respectively.

The anomeric methine ^{1}H signal at 5.50 ppm is observed to correlate with the C18 signal at 179.6 ppm in the gHMBC spectrum. We assign this anomeric ^{1}H signal to site 1' (on ring I):

$\delta_H = 5.50$ ppm and $\delta_C = 95.0$ ppm. The H1' signal at 5.50 ppm is observed to share a COSY cross peak with the midfield methine ^{1}H signal at 3.55 ppm. Unfortunately, six ^{1}H resonances are within 0.03 of 3.55 ppm, but the spin pair generating the signals labeled O is the best fit.

The most upfield of the four anomeric ^{1}H's at 4.79 ppm is observed to share a gHMBC cross peak with the C13 signal at 88.1 ppm. We assign the anomeric ^{1}H signal at 4.79 ppm to ring II's site 1″: $\delta_H = 4.79$ ppm and $\delta_C = 96.8$ ppm.

The H1' signal at 4.79 ppm has a single correlation in the COSY spectrum with what can only be a methine ^{1}H signal at 3.80 ppm. The only other ^{1}H signals near 3.80 ppm are from methylene groups which, being at the other end of the ^{1}H spin system, are not expected to couple most strongly to the anomeric ^{1}H. This allows us assign the ^{1}H/^{13}C spin pair labeled E in Table 5.3.3 to the 2″ site: $\delta_H = 3.80$ ppm and $\delta_C = 79.5$ ppm.

Because ring III is attached to ring II at the 2″ site, we can therefore attribute the gHMBC cross peak between the anomeric ^{1}H signal at 4.92 ppm and C2″ at 79.5 ppm to the H1‴-C2″ coupling, allowing us to assign the 4.92 ppm ^{1}H signal to ring III's site 1‴: $\delta_H = 4.92$ ppm and $\delta_C = 102.9$ ppm.

The last anomeric ^{1}H signal at 4.83 ppm is, because we have assigned all the other anomeric sites, assigned to site 1⁗ on ring IV. We write for site 1⁗: $\delta_H = 4.83$ ppm and $\delta_C = 103.1$ ppm.

Looking in the gHMBC spectrum, we observe that the H1⁗ signal at 4.83 ppm correlates most strongly with the label D ^{13}C signal at 86.1 ppm. Given that the anomeric ^{1}H's for the other rings are observed to correlate most strongly in the gHMBC spectrum across their attaching ether linkage, we assign the ^{13}C signal at 86.1 ppm to ring II's site 3″: $\delta_H = 3.93$ ppm and $\delta_C = 86.1$ ppm.

The ring IV anomeric ^{1}H signal (H1⁗) is also observed to correlate in the gHMBC spectrum with a ^{13}C signal near 77 ppm, which could correspond to labels H, I, J, or K. The H1⁗ signal also shares a well-resolved gHMBC cross peak with the label N ^{13}C signal at 74.3 ppm.

We are able to use the 120 ms mixing time TOCSY spectrum to provide some measure of clarity to the signals of the four spin system of the four sugar rings.

Using the labels in the first column of Table 5.3.3, we examine the 120 ms mixing time TOCSY spectrum and group the letters into the four spin systems for the four sugar rings. Each ring's spin system will have one anomeric methine, four garden-variety methines, and one methylene group (T, U, V, or W). Therefore each sugar ring's spin system must have five letters, one of which corresponds to a methylene group.

It is important to note that some of the HSQC ^{1}H/^{13}C signal pairs listed in Table 5.3.3 have been obtained from regions of the HSQC spectrum where the cross peaks overlap, casting into doubt some of the ^{1}H chemical shifts listed. In any case where we are including a letter corresponding to an overlapping ^{1}H resonance, we must be prepared to substitute other shift-similar possibilities as we work through the four sets of spins to find a self-consistent set of spins.

Our first pass through the 120 ms TOCSY spectrum yields the following sets of spins:

Ring I (1'-6'): 3.51 ppm (J), 3.55 ppm (O or H), 3.62 ppm (I and F), 3.78 ppm (V, cannot be E), 3.94 (V, cannot be D).

Ring II (1″-6″): 3.42 ppm (L), 3.57 ppm (S or K), 3.78 ppm (must be U and E, cannot be Q), 3.93 ppm (U and D).

Ring III (1‴-6‴): 3.33 ppm (M and P), 3.45 ppm (G), 3.55 ppm (H or O), 3.71 ppm (T), 3.93 ppm (T and/or D).

Ring IV (1⁗-6⁗): 3.44 ppm (N), 3.48 ppm (R), 3.58 ppm (K or S), 3.76 ppm (W and Q, cannot be E), 3.95 ppm (W, cannot be D).

Recall that we have already determined that the methines labeled D and E are on ring II at sites 3″ and 2″, respectively. We have also determined that the methine labeled N is on ring IV.

We must assume that M and P are on the same ring, if we are to believe that their signals are both present in the TOCSY spectrum. We invoke the same argument for the spins labeled I and F.

Because ring I has J, O or H, I, F, and V, we cannot consider Q as being a member of the ring I spin system. Ring II has L, S or K, E, D, and U, leaving no space for Q, again because each spin system can have only five letters. Ring III has no ^{1}H signal near the chemical shift of methine Q, and so Q must be included as part of the ring IV spin system.

There are multiple combinations we can create to arrive at four spin systems, each with five letters. When overlap creates ambiguity, we must examine the COSY and gHMBC spectra to find cross peaks that will help us eliminate impossible combinations. That is, even though the ^{1}H chemical shifts alone are not sufficient to allow us to separate the spins of the four sugar ring spin systems, we can use the gHMBC spectrum to determine which spin pairs belong to each spin system.

Our first point of uncertainty to resolve is whether label O (versus H) is on ring I or III. If methine O is part of the ring I spin system, then methine H must be part of the ring III spin system. Careful examination of the gHMBC spectrum shows that the ^{1}H signal at 3.62 or 3.63 ppm (I and/or F) correlates with the moderately resolved ^{13}C signal of methine O at 72.9 ppm. Therefore we are able to state that methine O is part of the ring I spin system and therefore that methine H is in the ring III spin system.

The second point of uncertainty that we will resolve is whether ring II or IV has methine S (versus K). The ^{1}H signal at 3.92 ppm (D, T, U, or V) is observed to correlate in the gHMBC spectrum with the well-resolved label S ^{13}C signal at 69.4 ppm. Because ring IV does not contain the methine labeled D, nor the methylenes labeled T, U, or V, we deduce that methine S must be part of the ring II spin system. The gHMBC cross peak we observe involves a ^{1}H from either D or U.

We can now write out the letter labels for our four sugar ring spin systems.

Ring I: F, I, J, O, V.
Ring II: D, E, L, S, U.
Ring III: G, H, M, P, T.
Ring IV: K, N, Q, R, W.

Having paired a single set of methylene signals with each sugar ring, we write for site 6′: δ_H = 3.775 & 3.94 ppm and δ_C = 61.6 ppm, for site 6″: δ_H = 3.78 & 3.92 ppm and δ_C = 61.8 ppm, for site 6‴: δ_H = 3.71 & 3.925 ppm and δ_C = 62.4 ppm, and for site 6⁗: δ_H = 3.77 & 3.96 ppm and δ_C = 61.6 ppm. It is now our task to order the methines from sugar rings sites 2 to 5, since we know the anomeric signals for the four site 1's and the methylene signals for the four site 6's.

We start with ring II, because we already have assigned NMR signals to sites 1″ (anomeric), 2″ (E), 3″ (D), and 6″(U). The ^{1}H signal for methine S at 3.58 ppm shares a gHMBC cross peak with C6″ at 61.8 ppm (label U). We assign the label S NMR signals to site 5″: δ_H=3.58 ppm and δ_C=69.4 ppm.

By the process of elimination, we can therefore assign the label L methine signals to site 4″: δ_H=3.43 ppm and δ_C=76.2 ppm. We can confirm that label L is correct for 4″ and S for 5″ because the 40 ms mixing time TOCSY spectrum indicates that the anomeric magnetization is passed more quickly to the label L ^{1}H signal at 3.43 ppm than to the label S ^{1}H signal at 3.58 ppm. Ring II's assignments are complete.

Examination of the COSY spectrum shows that the anomeric ^{1}H for ring I (H1') at 5.50 ppm couples to a ^{1}H signal at 3.55 ppm. We assign the 3.55 ppm ^{1}H signal to that of methine O, so for site 2': δ_H = 3.55 ppm and δ_C = 72.9 ppm. Although the methine labeled J is listed with a ^{1}H chemical shift of 3.51 ppm in Table 5.3.3, in the 120 ms TOCSY spectrum shown in Fig. 5.3.8, we can see that the ^{1}H signal of methine J appears to have a chemical shift that is closer to 3.50 ppm than to 3.51 ppm. This allows us to be confident that O is the correct label to assign to the 2' site, not J.

The ^{1}H signal at 3.62 or 3.63 ppm (I or F) is observed to share a gHMBC cross peak with the C2' signal at 72.9 ppm (label O). Overlap is too extensive in the gHMBC spectrum to determine if the H2' signal at 3.55 ppm shares a gHMBC cross peak with the ^{13}C signals of methine I versus F.

We now approach the sugar ring I spin system from its methylene terminus, for although the overlap between the ^{13}C signals of the Table 5.3.3 rows labeled V and W at 61.6 ppm is significant, we can, by excluding improbable correlations, obtain useful correlations. There is a gHMBC cross peak between at ^{1}H signal at 3.53 or 3.54 ppm (possibly labels H, O, J, R) and the ^{13}C chemical shift of 61.6 ppm (V and W). Because this locus in the gHMBC spectrum exhibits the only cross peak intensity involving two ^{13}C signals at 61.6 ppm, we can reasonably suppose that there are in fact two cross peaks at this position in the gHMBC spectrum, one involving the label V methylene group and one involving the label W methylene group. Of concern at this moment is the sugar ring I ^{1}H signals involved in generating cross peak intensity with the V label ^{13}C signal at 76.8 ppm (the other correlation is the label R ^{1}H signal to the C6''' signal on sugar ring IV). Because the label O ^{1}H signal is already assigned to site 2' and is not expected to correlate with the C6' signal, we therefore are only left with the label J methine signal as our selection for the 5' site of sugar ring I. The label R and H spins are not part of the sugar ring I spin system and are therefore excluded from consideration. We therefore assign the label J signals to site 5': δ_H = 3.51 ppm and δ_C = 76.84 ppm. We obtain a confirmation of this assignment by noting that, in the 40 ms TOCSY spectrum, anomeric ^{1}H magnetization is passed to the signals labeled I and F strongly at 3.62 and 3.63 ppm, and more weakly to the label J methine ^{1}H signal at 3.51 ppm.

Additionally, a gHMBC cross peak between the label J ^{1}H signal at 3.51 ppm and the label F ^{13}C signal at 77.6 ppm is observed. This places the label F methine group at the 4' site: δ_H = 3.62 ppm and δ_C = 77.6 ppm.

By the process of elimination we assign the label I methine group to site 3': δ_H = 3.63 ppm and δ_C = 77.0 ppm.

Moving on to ring III, we begin by noting a COSY cross peak shared between the anomeric site 1''' ^{1}H signal at 4.92 ppm and the most upfield of the midfield methine ^{1}H signals at 3.32 or 3.33 ppm. The 40 ms and 80 ms mixing time TOCSY spectra, when compared with the 120 ms mixing time TOCSY spectrum, show that the site 1''' anomeric ^{1}H magnetization is, for the shorter mixing time spectra, passed initially to the upfield side of the cross peak near 3.32/3 ppm. This allows us to assign the label M methine signals to site 2''': δ_H = 3.32 ppm and δ_C = 75.1 ppm.

We observe a gHMBC cross peak between the label H ^{1}H signal at 3.56 ppm and the well-resolved C2''' signal at 75.1 ppm. We assign the label H NMR signals to site 3''': δ_H = 3.56 ppm and δ_C = 77.0 ppm.

Approaching this spin system from its methylene terminus, we observe a gHMBC cross peak from the upfield H6''' at 3.71 ppm (label T) and the ^{13}C signal at 77.3 (label G), excluding from consideration the 77.0 ppm signal of label H which has already been assigned to site 3'''. We assign the label G methine NMR signals to site 5''': δ_H = 3.45 ppm and δ_C = 77.3 ppm.

By the process of elimination we assign the label P methine NMR signals to site 4′″: $\delta_H = 3.33$ ppm and $\delta_C = 71.2$ ppm. We obtain a weak confirmation of the 4′″ assignment by noting the gHMBC cross peak (weak because the cross peak lies in a crowded region of the spectrum) between H3′″ at 3.57 ppm and C4′″ at 71.2 ppm.

We will now finish the assignment of the NMR signal of Rebaudioside-A to its molecular sites by working through the cross peaks associated with sugar ring IV. The H1′‴ signal at 4.83 ppm shares a COSY cross peak with the label N ^{1}H signal at 3.44 ppm. We assign the label N ^{1}H and ^{13}C signal to site 2′‴: $\delta_H = 3.44$ ppm and $\delta_C = 74.3$ ppm.

Note that the signal of the anomeric ^{1}H of site 1′‴ shares two gHMBC cross peaks with a ^{13}C signal near 77 ppm which must be the label K ^{13}C signal at 76.77 ppm, and with the label R ^{13}C signal at 70.2 ppm. Also recall that the label R ^{1}H signal was observed to correlate in the gHMBC spectrum with the ^{13}C signal from site 6′‴ (label W). Because the anomeric ^{1}H at site 1′‴ has a signal that participates in more gHMBC cross peaks than its other anomeric compatriots, we assume that sugar ring IV is more rigid than the other sugar rings. Therefore it is reasonable that the anomeric ^{1}H at site 1′‴ is able to couple with spins on the other side of the sugar ring IV ether linkage. This allows us to assign the label R signals to site 5′‴: $\delta_H = 3.49$ ppm and $\delta_C = 70.2$ ppm. The label K NMR signals are assigned to site 3′‴: $\delta_H = 3.58$ ppm and $\delta_C = 76.77$ ppm.

We do not observe a cross peak in the 40 ms TOCSY spectrum between the H1′‴ signal and the label Q ^{1}H signal at 3.75 ppm. We therefore assign the label Q NMR signals to site 4′‴: $\delta_H = 3.75$ ppm and $\delta_C = 70.5$ ppm. We must suppose that the rigidity of the sugar ring IV is sufficient to allow H1′‴ to couple directly to H5′‴ across the sugar ring IV ether linkage, thereby allowing the H1′‴ signal to correlate in the 40 ms TOCSY spectrum with the H5′‴ signal (label R), but not with the H4′‴ signal (label Q). That is, there is a small $^4J_{HH}$ between H1′‴ and H5′‴ across the sugar ring IV ether linkage that allows the anomeric H1′‴ magnetization to sneak around to the other end of the sugar ring IV spin system, such that the 4′‴ site is the last methine to receive the H1′‴ magnetization during the mixing period of the TOCSY NMR experiment.

5.4 B3-MONOHYDROXY GLYCEROL DIBIPHYTANYL GLYCEROL TETRAETHER IN BENZENE-D$_6$

A graduate student studying at the University of Bremen named Xiaolei Liu was visiting Professor Roger Summons in the Department of Earth, Atmospheric and Planetary Sciences at the Massachusetts Institute of Technology in late 2010 when he approached me with 400 micrograms of a natural product with a formula weight of 1318.36 g mol^{-1} (Liu et al., 2012). The material was B3-Monohydroxy Glycerol Dibiphytanyl Glycerol Tetraether and is abbreviated as GDGT-OH. The structure of GDGT-OH is shown in Fig. 5.4.1.

Interestingly, those tasked with originally numbering the sites of the GDGT molecule (1) used letter prefixes for different portions of the molecule; (2) assigned a greater priority to the hydroxyl bearing glycerol carbons than to those carbons participating in ether linkages; (3) denoted the site of hydroxyl functionalization as being on the B chain (sites with a letter B prefix) when, by symmetry, the site of hydroxylation could have been A3′. Rather than renumber the molecule, the author elects to preserve the numbering system already extant in literature.

The sample had been obtained by painstakingly extracting this natural product (is there any other type of natural product extraction?) from marine sediment obtained from core

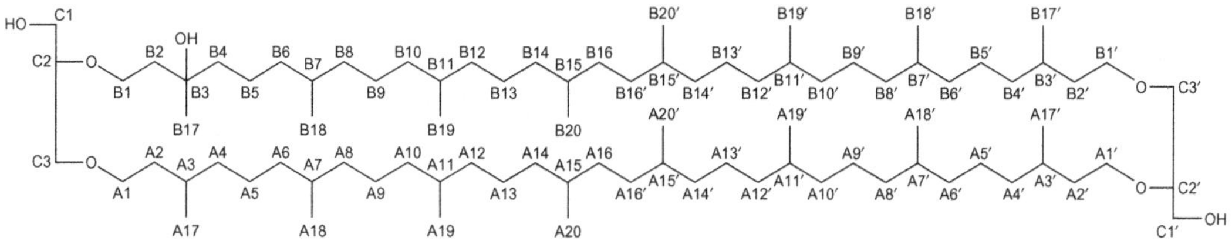

FIG. 5.4.1 The structure of B3-monohydroxy glycerol dibiphytanyl glycerol tetraether.

samples of the floor of the North Atlantic Ocean. The sample was dissolved in 40 microliters of benzene-d_6 and a number of NMR experiments were carried out using a Varian/Agilent nanoprobe, which spins a 40 μL sample at about 2 kHz around an axis inclined at the magic angle with respect to the direction of the applied magnetic field. Probe performance was not optimal, as the probe had been used heavily in the past by chemists in the group of Peter Seeberger when he was at MIT in the early 2000s to examine materials bound to the surface of gel beads. The spectra from the sample were, considering the number of moles of material (4.0×10^{-4} g/1318.36 g mol^{-1} = 300 nmol), of sufficient quality as to be useful. Dissolution of the 300 nmol of material into a 40 μL sample volume gave a solution with a concentration of 7.6 mM, assuming accurate mass, high purity, and quantitative transfer of sample solution into the sample rotor. Some of the spectra that were collected were done with a rotor that was not spinning freely, and so chatter-induced artifacts appear in some of the spectra. It is unfortunate that some of the spectra collected from this precious sample suffered the deleterious effects of a poorly engaged air bearing. The small amount of sample and the fact that the NMR probe used was an inverse probe (optimized for ^{1}H observation but not for ^{13}C observation) also prevented the collection of a 1-D ^{13}C NMR spectrum.

Despite the lack of a 1-D ^{13}C NMR spectrum, we still rely on the dispersion of ^{13}C signals as our main tool in unraveling the panoply of sites and signals in our molecule. The ^{1}H NMR signals are, in total, simply too numerous and, because of the multiplet widths along the chemical shift axis, overlapped too extensively to allow us to make assignments without the ^{13}C signal dispersion present in the system. To glean tidbits of useful information stepwise from this complex molecule, we use 2-D methods to introduce additional dispersion and thereby to metaphorically chip away at the problem, hopefully sculpting something meaningful in the end. This problem represents a change in course or policy: We are crossing the Rubicon from comfortable and discrete site-to-signal assignments into the world of polymers, with a high degree of homology between repeat units where the shoulders, edges, and tails associated with the main signals can become the carrier of nuanced import. Put another way, we will have to first make sense of the main signals and, having established their context, we will then gain understanding by observing how deviations from the dominant chemical environment perturb the chemical shifts of the observed signals.

The 1-D ^{1}H NMR spectrum of GDGT-OH in benzene-d_6 is shown in Fig. 5.4.2. As we might expect from a molecule with 172 hydrogens, there is signal overlap along the ^{1}H chemical shift axis. The 2-D ^{1}H-^{1}H COSY NMR spectrum appears in Fig. 5.4.3. The 2-D ^{1}H-^{13}C HSQC NMR spectrum is shown in Figs. 5.4.4–5.4.6. The 2-D ^{1}H-^{13}C gHMBC NMR spectrum is found in Fig. 5.4.7.

Examination of the structure of GDGT-OH enables us to comprehend what will be entailed in the task of assignment. We begin by describing the GDGT molecule without the extra hydroxyl group that appends the "–OH" to its abbreviation. The central portion of the GDGT molecule consists of two isoprene chains, each chain of which is composed of four head-to-tail

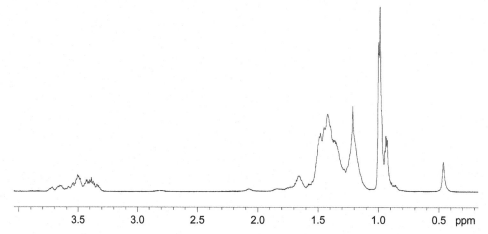

FIG. 5.4.2   The 1-D ^{1}H NMR spectrum of B3-monohydroxy glycerol dibiphytanyl glycerol tetraether in benzene-d_6.

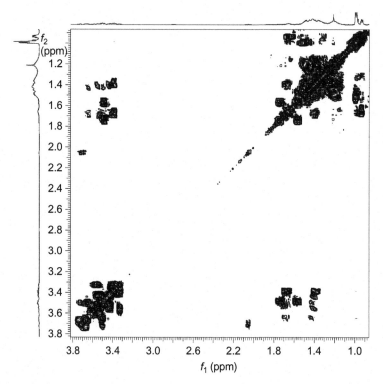

FIG. 5.4.3   The 2-D ^{1}H-^{1}H COSY NMR spectrum of B3-monohydroxy glycerol dibiphytanyl glycerol tetraether in benzene-d_6.

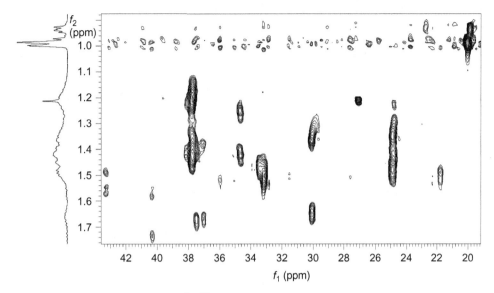

FIG. 5.4.4   The upfield portion of the 2-D ^{1}H-^{13}C HSQC NMR spectrum of B3-monohydroxy glycerol dibiphytanyl glycerol tetraether in benzene-d_6.

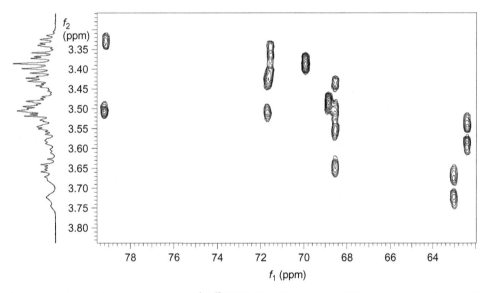

FIG. 5.4.5   The downfield portion of the 2-D ^{1}H-^{13}C HSQC NMR spectrum of B3-monohydroxy glycerol dibiphytanyl glycerol tetraether in benzene-d_6.

isoprene repeat units joined head-to-head (the methyl group of each isoprene unit determines which end is the head). Each of these chains is said to be a tetraterpene, in that it has eight isoprene units contributing 40 carbon atoms. Because GDGT is not a hydrocarbon, it is said to be a terpenoid rather than just a terpene, although adherence to the terpene/terpenoid distinction is not always rigorous in the literature. The ends of both tetraterpene chains are

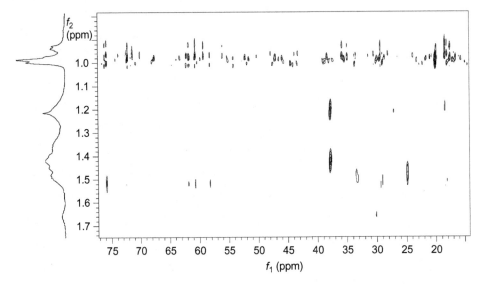

FIG. 5.4.6 An additional portion of the 2-D ^{1}H-^{13}C HSQC NMR spectrum of B3-monohydroxy glycerol dibiphy-tanyl glycerol tetraether in benzene-d_6, showing correlations between upfield ^{1}H signals and midfield ^{13}C signals.

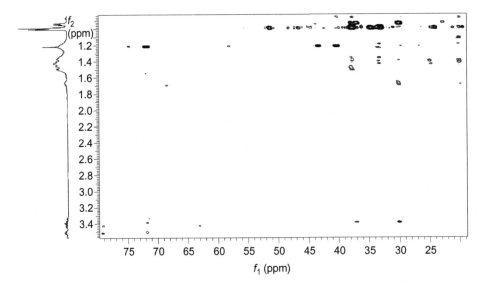

FIG. 5.4.7 The 2-D ^{1}H-^{13}C gHMBC NMR spectrum of B3-monohydroxy glycerol dibiphytanyl glycerol tetraether in benzene-d_6.

bound via ether linkages (hence the *tetraether* in the name) to two glycerol residues at adjacent sites, i.e., the glycerol ends are not 1,3-substituted but rather 1,2-substituted. The ends of a given isoprene chain are believed to be bound to different glycerol positions: The top chain is attached to C2 of the left glycerol and C3′ of the right glycerol. The attachment configuration of the isoprene chains to the glycerol ends gives the GDGT molecule a c_2 symmetry axis. All of the isoprene residues are chiral, however, so there is no mirror plane in GDGT.

Once a hydroxyl group is added to the carbon three bonds from the ether linkage of the C2-position of a glycerol residue, all symmetry elements vanish and we have GDGT-OH. In practice, the addition of a single hydroxyl group to an elongated C_{40} hydrocarbon has little effect on most of the resonances of the atoms that are well removed from the attachment point of the extra hydroxyl group.

Much of the interpretation of the spectra involves identifying main peaks associated with the unoxidized isoprene repeat units. It would also be unfair to not admit that the author, lacking extensive knowledge of the NMR behavior of larger terpenoids, resorted to ^{13}C chemical shift prediction to provide guidance and perspective in mentally digesting the scope and complexity of the molecule GDGT-OH.

We will forego the prediction of ^1H resonance multiplicities for GDGT-OH, other than to note that the B17 ^1H signal is expected to be a singlet because the B17 methyl is bound to a nonprotonated ^{13}C, while the ^1H signal of every other methyl group is expected to be a doublet—split by its neighboring methine ^1H. At the left and right sides of the molecule, as it is shown in Fig. 5.4.1, we find eleven carbons bound to oxygen: C1, C2, C3, A1, B1, B3, C1', C2', C3', A1', B1'. Only B3 is nonprotonated, so we expect to find 10 midfield sets of cross peaks in the HSQC spectrum corresponding to the signals of the 10 protonated carbon atoms alpha to oxygen. Of these 10 sets of signals, only two will be from methine groups (C2 and C2') and the remainder will be from methylenes.

This project began with the isolation of a molecule with an empirical formula of $C_{86}H_{172}O_7$, which was the empirical formula for GDGT plus one oxygen. For the unoxidized tetraterpene chains, the major methyl ^1H signal is found at 0.95 ppm. The main clue as to the location of the oxygen in the molecule is the observation of a downfield-shifted methyl singlet at 1.21 ppm, suggesting that a hydroxyl group has replaced the hydrogen atom of a methine group in one of the isoprene repeat units. Determination of which isoprene unit is based on the recognition that functionalization near the central portions of the chains would not appreciably effect the chemical shifts of one but not the other of the glycerol ends of the molecule. If the OH-for-H substitution occurred in the middle of the tetraterpene chain, we would expect to see identical glycerol residue signals.

Placement of the seventh oxygen on an isoprene unit next to one of the glycerol residues perturbs the chemical shifts of the glycerol residue signals, enabling us to assert that the site of oxidation is on the first isoprene repeat unit. If the new hydroxyl group were located at site B7 on the second isoprene unit, the oxygen on site B7 would be six bonds from B1, too far for the spins in between to experience much of an effect from both sites. In other words, if the new hydroxyl is on site B3, then the site B2 signals will experience shift perturbations from the oxygen on the other side of B1 and also from the hydroxyl on B3. If the hydroxyl is six bonds away, it is unlikely we will identify any commonly perturbed signals, meaning that position information would not be available.

If we consider each isoprene unit in the tetraterpene chains, we are able to anticipate the general appearance of the spectra. At the ends of the chains, we have isoprene units one and eight, both of which are connected via an ether linkage to the glycerol residues ends. We therefore expect the chain ends to generate resonances that are shifted downfield from their midchain counterparts. At the midpoint of each tetraterpene chain, the head-to-head coupling of isoprene units imparts to the signals of isoprene units four and five slightly different chemical shifts than what is observed of the signals of units two, three, six, and seven—signals of the site 16 methylenes at the chain centers should especially possess a unique ^{13}C chemical shift, reflecting their unique chemical environment. The main ^{1}H and ^{13}C NMR signals, therefore, will be associated with isoprene units two, three, six, and seven. Units four and five are effected on their head sections (e.g., sites B15, B16, B20) owing to the head-to-head connection of units four and five. The chemical shifts of the spins in units one and eight are perturbed at the tail of the isoprene unit (e.g., sites A1, A2, and A3).

Fig. 5.4.8 shows the predicted ^{13}C chemical shifts of the signals from an isoprene unit in the middle of a long terpene chain. Chemical shift prediction software indicates that the site i methylene, which is bracketed by two other methylene groups, will generate a signal with a ^{13}C chemical shift of 25 ppm. For sites ii and iv, with a methylene on one side and a methine on the other, their ^{13}C's are expected to generate signals with a chemical shift of 38 ppm. The site iii methine, with two attached methylenes and a pendant methyl group, is expected to generate a ^{13}C resonance at 33 ppm. The methyl group at site v should produce a ^{13}C signal at 20 ppm. Table 5.4.1 lists the GDGT-OH sites expected to fall into one of these four main regions of the ^{13}C chemical shift axis.

The information in Table 5.4.1 allows us to anticipate that we should observe HSQC signals from the four distinct carbon environments (i, ii/iv, iii, and v). Superimposed upon the dominant four sets of signals from the terpene chains, we expect to observe the signals from those sites in the middle of the chains, where the head-to-head isoprene linkages are found, and near the ether linkages at the tetraterpene chain ends. Table 5.4.2 lists which sites of GDGT-OH are expected to generate signals that deviate, and how the chemical shifts of these signals will deviate (based on chemical shift prediction software) from the dominant site/signal pairings in Table 5.4.1.

If we examine the upfield portion of the HSQC spectrum, we are able to identify the dominant cross peaks associated with the sites listed in Table 5.4.1. Moving from left to right (f_1 is horizontal), the first major set of HSQC signals we encounter are two broad ^{1}H signals at 1.20

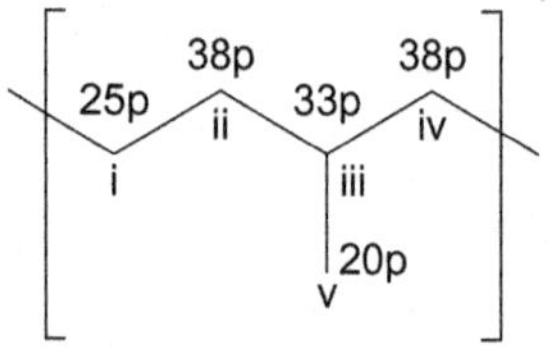

FIG. 5.4.8 The structure of the isoprenoidal repeat unit with the five carbon sites bearing labels showing the predicted chemical shifts of its ^{13}C signals.

TABLE 5.4.1 Sites of GDGT-OH Grouped by Isoprene Repeat Unit Chemical Environment

Repeat Unit Site (Fig. 5.4.8)	^{13}C Chemical Shift (ppm)	Corresponding GDGT-OH Sites
i	25	A5, A9, A13, A5′, A9′, A13′, B9, B13, B5′, B9′, B13′
ii, iv	38	A4, A6, A8, A10, A12, A14, A4′, A6′, A8′, A10′, A12′, A14′, B6, B8, B10, B12, B14, B4′, B6′, B8′, B10′, B12′, B14′
iii	33	A7, A11, A7′, A11′, B7, B11, B7′, B11′
v	20	A17, A18, A19, A20, A17′, A18′, A19′, A20′, B18, B19, B20, B17′, B18′, B19′, B20′

TABLE 5.4.2   Sites of GDGT-OH Expected to Generate Signals that Deviate from the Dominant Chemical Environment

Repeat Unit Site (Fig. 5.4.1)	^{13}C Chemical Shift (ppm)	Corresponding GDGT-OH Sites
i	71	A1, B1'
i	70	A1'
i	67	B1
i	22	B5
ii	38	A2, A2', B2'
ii	45	B2
iii	30	A3, A3', B3'
iii	75	B3
iii	32	A15, A15', B15, B15'
iv	42	B4
iv	34	A16, A16', B16, B16'
v	26	B17

& 1.43 ppm that correlate with ^{13}C signal at 37.7 ppm. We assign these signals to the methylene sites ii and iv positions in the isoprene repeat unit. We write for sites ii/iv: $\delta_H = 1.20$ & 1.43 ppm and $\delta_C = 37.7$ ppm (sites A4, A6, ..., B14'). The next set of signals we encounter are the methine HSQC cross peaks which are opposite in sign to the methylene cross peaks. The dominant methine HSQC signal intensity is found at a ^{1}H chemical shift of 1.50 ppm and a ^{13}C shift of 33.1 ppm. We assign this correlation to site iii: $\delta_H = 1.50$ ppm and $\delta_C = 33.1$ ppm. Proceeding to lower ^{13}C chemical shift values, we find two more of the strongest HSQC cross peaks at ^{1}H chemical shifts of 1.35 & 1.48 ppm and the ^{13}C shift of 24.9 ppm. We assign these signals to site i: $\delta_H = 1.35$ & 1.48 and $\delta_C = 24.9$ ppm. Finally we observe the strongest methyl HSQC cross peak at a ^{1}H shift of 0.99 ppm and a ^{13}C shift of 20.0 ppm. We assign these signals to site v: $\delta_H = 0.99$ ppm and $\delta_C = 20.0$ ppm.

Having identified the dominant HSQC signals associated with eight or more molecular sites (cross peaks whose intensity results from at least eight sets of ^{1}H/^{13}C signal contributors), we next identify the signals from the tetraterpene chains associated with the middle of the chains, i.e., those sites whose shifts are perturbed by the head-to-head joining of isoprene units four and five. These midchain HSQC signals should possess half the intensity or less of the dominant peaks, because only four molecular sites contribute to these midchain NMR signals. Searching near a ^{13}C chemical shift of 34 ppm, we observe two well-resolved HSQC cross peaks from ^{1}H's with shifts of 1.25 & 1.42 ppm to a ^{13}C signal at 34.7 ppm. We assign these signals to sites A16, A16', B16, and B16': $\delta_H = 1.25$ & 1.42 ppm and $\delta_C = 34.7$ ppm.

The methine HSQC cross peaks associated with sites A15, A15', B15, and B15' are found as a shoulder of the main site iii HSQC signal. We observe an HSQC cross peak with a sign consistent with a methine (or a methyl) group at the ^{1}H chemical shift of 1.47 ppm and a ^{13}C shift of 33.4 ppm, with a ^{1}H shift that differs by 0.03 ppm from the main site iii ^{1}H shift of 1.50 ppm and a ^{13}C shift just 0.3 ppm greater than the main site iii shift of 33.1 ppm. We write for sites A15, A15', B15, B15': $\delta_H = 1.47$ ppm and $\delta_C = 33.4$ ppm. In the gHMBC spectrum, we observe a pair of cross peaks between the site 16 ^{1}H signals at 1.25 & 1.42 ppm and the site 15 ^{13}C signals at 33.4 ppm, consistent with our assignment of sites 15 and 16.

Just as we observe the midchain methine HSQC signal as a shoulder to the main site iii methine signal, so, too, do we observe a methyl HSQC cross peak very near to the dominant methyl HSQC signal. We observe a ^{1}H signal at 0.98 ppm that correlates with a ^{13}C signal at 19.6 ppm. Another methyl-spawned HSQC signal is also observed nearby at a ^{1}H shift of 0.94 ppm and a ^{13}C shift of 19.8 ppm. If we examine the 1-D ^{1}H NMR spectrum, we can see that the ^{1}H signal at 0.94 ppm is not a clean doublet, which therefore disqualifies this ^{1}H signal from consideration for assignment to the sites A20, A20', B20, and B20'—all of which are expected to be isochronous or nearly so. The other HSQC methyl shoulder signal, however, appears in the 1-D ^{1}H spectrum as a clean doublet. It is noted that only one of the legs of the doublet is resolved, while the downfield leg overlaps with the upfield leg of the dominant methyl signal associated with the site v methyl groups. Nonetheless, the obviously more complex nature of the 0.94 ppm ^{1}H methyl signal suggests a greater variation of chemical environment than we can reasonably attribute to the midchain methyl groups. We assign the less complex ^{1}H signal to sites A20, A20', B20, B20': $\delta_H = 0.98$ ppm and $\delta_C = 19.6$ ppm. The other set of more ^{1}H-environment-varied HSQC signals we attribute to sites A17, A17', and B17': $\delta_H = 0.94$ ppm and $\delta_C = 19.8$ ppm. Note that we had originally expected to find the NMR signals from sites A17, A17', and B17' lumped in with the dominant methyl signal, so we are pleasantly surprised to obtain this additional assignment, however tenuous and unverified it may be. Closer examination of the ^{1}H signals near 0.94 reveals what appears to be three overlapping doublets, with two of the doublets nearly superimposed. The observation of three doublets near 0.94 ppm is consistent with our notion that this is the ^{1}H signal from *three* methyl sites (A17, A17', and B17'). We might even argue that the overlapping ^{1}H signals (slightly more upfield) are those from sites A17 and B17'.

We observe the HSQC signal of methine groups whose intensity is greater than that from a single site at a ^{1}H shift of 1.65 ppm and a ^{13}C shift of 30.1 ppm. This HSQC correlation is strong, but less intense than that of the main methine signal. We assign this signal to the sites A3, A3', and B3': $\delta_H = 1.65$ ppm and $\delta_C = 30.1$ ppm. This is an example of how chemical shifts often move counter intuitively upfield when an electronegative atom is several bonds distant—in this case the ether oxygen pulls the alpha and beta methylenes (e.g., A1 and A2) downfield, but the gamma methine has a ^{13}C chemical shift that is upfield (one ppm lower) from the chemical shift of the main methine signal. Our assignment of sites A3, A3', and B3' and sites A17, A17', and B17' are consistent: The ^{1}H signal of the A17, A17', and B17' methyl groups at 0.98 ppm correlates in the gHMBC spectrum with the ^{13}C signal of sites A3, A3', and B3' at 30.1 ppm.

In this upfield region of the HSQC spectrum we have exhausted all of the chemical environment redundancies expected to generate correlations involving multiple sites. We go on to tabulate the remaining signals that are associated with individual sites in this 86-carbon molecule. Table 5.4.3 lists the remaining unassigned signals observed in the upfield portion of the HSQC spectrum.

The ^{1}H signal of the B17 methyl group (or whichever other methyl is attached to the hydroxylated former methine carbon) is a clear singlet at 1.21 ppm. We write for site B17: $\delta_H = 1.21$ ppm and $\delta_C = 27.1$ ppm.

TABLE 5.4.3   Upfield HSQC Signals of GDGT-OH and the Molecular Site Likely Associated Therewith

[1]H Shift (ppm)	[13]C Shift (ppm)	Possible Associated Site
1.49 & 1.56	43.3	B2
1.57 & 1.73	40.3	B4
1.22 & 1.42	37.8	B6
1.41[a] & 1.68	37.5	A2, B2′
1.38 & 1.67	37.0	A2′
1.34	30.1	Methylene impurity
1.21	27.1	B17
1.49	21.8	B5
0.86	19.7	Methyl impurity

[a]*Overlap in the HSQC makes this chemical shift difficult to ascertain.*

In the gHMBC spectrum, the ^{1}H signal from site B17 participates in three cross peaks. The B17 ^{1}H signal correlates most strongly with a ^{13}C signal at 71.7 ppm. This ^{13}C shift is assigned to site B3, which is expected to have a chemical shift of 75 ppm. The B17 ^{1}H signals also correlate with ^{13}C resonances at 43.3 and 40.3 ppm, consistent with our belief that these 40-something ppm ^{13}C shifts are those from sites B2 and B4. In the gHMBC spectrum we also observe a correlation between the ^{1}H chemical shift of 1.56 ppm and the B3 ^{13}C resonance at 71.7 ppm. It is unfortunate that the nonprotonated B3 site has a ^{13}C chemical shift of 71.7 ppm—as this is the same shift that we observe for one of our midfield HSQC methylene ^{13}C signals (see later).

The gHMBC spectrum's strongest cross peak correlates the main methyl ^{1}H signal of site v at 0.99 ppm and the ^{13}C signals at 37.7 ppm assigned to the methylenes groups of sites ii and iv. The main methyl ^{1}H signal of site v is also observed to correlate strongly in the gHMBC spectrum with the methine ^{13}C signal of site iii at 33.1 ppm, and more moderately with the methylene ^{13}C resonance of site i at 24.9 ppm. The main methyl ^{1}H signal also shares a gHMBC cross peak with the ^{13}C signal we assigned to the site 16 methylenes at 34.7 ppm. Because this cross peak involves site 16 at the terpene chain midpoint, we expect to only have f_2 (^{1}H) contributions from the site 20 methyl groups at 0.98 ppm, we note that the cross peak is found centered at the ^{1}H shifts of 0.99 ppm, not the 0.98 ppm shift we had earlier supposed might be from the clean site 20 doublets.

We now proceed to the midfield portion of the HSQC spectrum, where we expect to observe the NMR signals of sites A1, A1′, B1, B1′, C1, C1′, C2, C2′, C3, and C3′. Our ^{13}C chemical shift predictions for the glycerol residue ends of GDGT-OH are such that we expect to observe C1/C1′ signals with chemical shifts of 62 ppm, C2/C2′ signals with shifts of 80 ppm, and C3/C3′ signals with shifts of 71 ppm. Here we anticipate observing eight methylene signal pairs and two methine correlations. We are happy to observe two methine cross peaks and eight pairs of methylene cross peaks in the midfield region of our HSQC spectrum. We measure the shifts of the cross peaks, placing the measured values into Table 5.4.4.

TABLE 5.4.4 Midfield HSQC Signals of GDGT-OH and Possible Molecular Sites Associated Therewith

Label	^{1}H Shift (ppm)	^{13}C Shift (ppm)	Possible Associated Site
a	3.51	79.2	C2 or C2'
b	3.33	79.1	C2' or C2
c	3.43 & 3.51	71.7	C3 or C3'
d	3.35 & 3.41	71.5	C3' or C3
e	3.37 & 3.39	69.9	A1, A1', B1, or B1'
f	3.48 & 3.50	68.8	A1, A1', B1, or B1'
g,h	3.44, 3.50, 3.56, 3.64	68.5	A1, A1', B1, or B1'
i	3.67 & 3.72	63.0	C1 or C1'
j	3.54 & 3.59	62.4	C1' or C1

Our task is now to pair these signals with one another, especially across the ether linkages. We will use the COSY and the gHMBC spectra for this endeavor. We will first identify the A1/A1'/B1/B1' to A2/A2'/B2/B2' COSY connections. Following that, we will identify the members of the glycerol residue spin systems. Lastly, we will use information in the gHMBC spectrum to link the terpene and glycerol spin systems unambiguously.

The ^{1}H signals from methylenes and methine groups alpha to oxygen (from sites A1, A1', B1, B1', C1, C1', C2, C2', C3, and C3') are all found in the midfield chemical shift region between 3.3 and 3.8 ppm. The ^{1}H signals of the methylenes beta to oxygen (from sites A2, A2', B2, and B2') are expected to correlate with the oxygen-alpha-methylene ^{1}H signals from A1, A1', B1, and B1' only.

The most downfield of the oxygen-beta ^{1}H signals at 1.57 & 1.73 ppm is observed to correlate in the COSY spectrum with the oxygen-alpha ^{1}H's at 3.48 & 3.50 ppm (label f). We had originally supposed that the ^{1}H signals at 1.57 & 1.73 ppm were those associated with site B4 because the ^{13}C chemical shift of the carbon atom bearing the ^{1}H's with shifts of 1.57 & 1.73 ppm was most consistent with the predicted ^{13}C shift of site B4. However, we must, in light of the COSY correlation, change our view of this methylene group's identity: We are now inclined to attribute the 1.57 & 1.73 ppm ^{1}H signals to site B2. We write the site B2: $\delta_H = 1.57$ & 1.73 ppm and $\delta_C = 40.3$ ppm. We are also able to assign site B1 based on the aforementioned COSY correlation, so for site B1: $\delta_H = 3.48$ & 3.50 ppm and $\delta_C = 68.8$ ppm (label f).

Based on the gHMBC correlation of a methylene ^{13}C signal at 43.3 ppm to the B17 methyl ^{1}H signal at 1.21 ppm that we touched upon previously, we assign the other B17-correlated methylene ^{13}C signal to site B4: $\delta_H = 1.49$ & 1.56 ppm and $\delta_C = 43.3$ ppm.

We assign the most upfield of all observed methylene ^{13}C signals to site B5: $\delta_H = 1.49$ & 1.53 ppm and $\delta_C = 21.8$ ppm. We also take this opportunity to assign the methylene NMR signals found at a ^{13}C chemical shift of 37.8 ppm—slightly downfield from the major sites ii and iv signal at 37.5 ppm—to the B6 site: $\delta_H = 1.22$ & 1.42 ppm and $\delta_C = 37.8$ ppm.

Another correlation observed in the COSY spectrum is between the signals of the oxygen-beta ^{1}H's at 1.38 & 1.67 ppm and the oxygen-alpha ^{1}H signals at 3.38 & 3.40 ppm (label e).

A third series of COSY cross peaks is found to connect the oxygen-beta ^{1}H signals at 1.38 & 1.67 ppm to the oxygen-alpha ^{1}H signals & 3.56 ppm (label g,h). Although we can clearly observe in the COSY spectrum only two of the cross peaks between the oxygen-beta ^{1}H signals at 1.41 & 1.68 ppm and the last oxygen-alpha ^{1}H signal at 3.65 ppm, we know from the HSQC spectrum (shifts listed in third from bottom of Table 5.4.4) that the other ^{1}H signal from the oxygen-alpha methylene group must be found at 3.50 ppm (label g,h). We have just paired the ^{1}H signals from the ends of the terpene chains and sorted the label g ^{1}H signals from those of label h.

The glycerol ends of GDGT-OH must therefore include the signals list in Table 5.4.4 rows with the following letter labels: a, b, c, d, i, and j. In the COSY spectrum, we observe a correlation between the most downfield of the label i ^{1}H signals at 3.72 ppm and the label a methine signal at 3.51 ppm. Therefore the label a and label i ^{1}H's are in the same spin system. The label b ^{1}H signal at 3.33 ppm is observed to share COSY cross peaks with ^{1}H signals at 3.54 ppm and 3.41 ppm. The 3.54 ppm signal can only be from the label j ^{1}H because other ^{1}H's with similar shifts (labels g,h) have already been grouped with the terpene chain ends. The 3.41 ppm signal must be from one of the label d ^{1}H's, because the label e signals have already been determined to be from a methylene group that is part of the terpene chains. We group the glycerol ends into the following two spin systems whose signals contain these sets of labels: b, d, j, and a, c, i.

We observe a gHMBC cross peak between the label b ^{1}H signal at 3.33 ppm and the label d ^{13}C signal at 71.5 ppm, thereby confirming that the sites associated with the signals labeled b and d are part of the same glycerol residue. We also observe a gHMBC cross peak between the label a ^{1}H signal at 3.51 ppm and the label c ^{13}C signal 71.7 ppm, also consistent with our placement of the spins generating the signals labeled a and c on the same glycerol residue.

From the ^{13}C shifts, with the assistance of our ^{13}C chemical shift predictions, we can write that the label a/b signals are from C2 and C2′, the label d and e signals are from label C3 and C3′, and the label i and j signals are from C1 and C1′.

In the gHMBC spectrum, the 3.33 ppm ^{1}H signal (label b) is observed to correlate with the site B1 ^{13}C signal with a chemical shift of 68.7 ppm (label f). We assign the 3.33 ppm ^{1}H signal to site C2: $\delta_H = 3.33$ ppm and $\delta_C = 79.1$ ppm. We can therefore assign all of the remaining glycerol NMR signals in the unprimed glycerol spin system. We assign the label j NMR signals to site C1: $\delta_H = 3.54$ & 3.59 ppm and $\delta_C = 62.4$ ppm. We assign the label d NMR signals to site C3: $\delta_H = 3.36$ & 3.41 ppm and $\delta_C = 71.5$ ppm.

By the process of elimination, we also are able to assign the NMR signals of the primed glycerol residue sites. We write for site C1′: $\delta_H = 3.67$ & 3.72 ppm and $\delta_C = 63.0$ ppm (label i). For site C2′: $\delta_H = 3.51$ ppm and $\delta_C = 79.2$ ppm (label a), and for site C3′: $\delta_H = 3.43$ & 3.51 ppm and $\delta_C = 71.7$ ppm (label c).

Although we do not observe many gHMBC cross peaks involving spins whose couplings span the glycerol-to-terpene-chain ether linkages, we do observe a gHMBC correlation between the label e ^{1}H signal at 3.39 ppm and the site C3′ ^{13}C signal at 71.7 ppm (label c). We assign the label e ^{1}H signal at 3.39 ppm to site B1′: $\delta_H = 3.37$ & 3.39 ppm and $\delta_C = 69.9$ ppm.

We also observe a second gHMBC cross peak involving the signals of sites B1′ and C3′ that confirms the placement of these two groups across a common ether linkage: The ^{1}H signal of site C3′ at 3.43 ppm (label c) shares a gHMBC cross peak with the B1′ ^{13}C signal at 69.9 ppm (label e). The site C3′ ^{1}H signal at 3.43 ppm (label c) also correlates in the gHMBC spectrum with the C1′ ^{13}C resonance at 63.0 ppm (label i)

The label e signals, which we now know belong to site B1′, were earlier noted as correlating in the COSY spectrum with oxygen-beta ^{1}H's with shifts of 1.38 & 1.67 ppm. We assign the oxygen-beta ^{1}H signals to site B2′: $\delta_H = 1.38$ & 1.67 ppm and $\delta_C = 37.0$ ppm.

If the 71.7 ppm ^{13}C signal involved in the gHMBC cross peak with the label e ^{1}H signal were from site B1, the label e ^{1}H signal would also be expected to correlate with the ^{13}C signal of site B2 at 40.3 ppm (it does not).

The site B1′ ^{1}H signal at 3.39 ppm is also observed to correlate in the gHMBC spectrum with the ^{13}C signal of site B2′ at 37.0 ppm and with a methine ^{13}C resonance at 30.1 ppm, which is attributed to site B3′ but could just as easily be from sites A3 or A3′. What is important is that the B1′ ^{1}H signal correlates with a ^{13}C signal that is clearly not that of site B3 (at 71.7 ppm). We therefore know that the ^{1}H signal at 3.39 ppm cannot be that of site B1, which is not close to a methine ^{13}C with a chemical shift of 30.1 ppm.

We are now faced with the task of assigning the NMR signals of sites A1, A2, A1′, and A2′. Lacking any gHMBC correlations, we can only assign the more downfield ^{13}C chemical shift to site A1 based on the ^{13}C predictions. The chemical shift of the site A1 ^{13}C was predicted to be 70.9 ppm, while that of site A1′ was predicted to be 69.2 ppm. Because the two oxygen-alpha ^{13}C shifts are the same (68.5 ppm), we assign, based only on a chemical shift alternation pattern we have already observed at other sites in this molecule, the NMR signals of the methylene group with the more upfield ^{1}H signals to site A1: $\delta_H = 3.44$ & 3.56 ppm and $\delta_C = 68.5$ ppm. Recalling the earlier COSY correlations we established between oxygen-beta and oxygen-alpha methylene ^{1}H signals, we write for site A2: $\delta_H = 1.41$ & 1.68 ppm and $\delta_C = 37.5$ ppm. By the process of elimination we write for site A1′: $\delta_H = 3.50$ & 3.64 ppm and $\delta_C = 68.5$ ppm, and for site A2′: $\delta_H = 1.41$ & 1.68 ppm and $\delta_C = 37.5$ ppm.

References

Berger, S., Braun, S., 2004. 200 and More NMR Experiments: A Practical Course. Wiley-VCH, Weinheim, Germany, ISBN: 3527310673. 854 pp.

Brevard, C., Granger, P., 1981. Handbook of Higher Resolution Multinuclear NMR. Wiley-Interscience, New York, ISBN: 0471063231. 229 pp.

Chopade, P.D., Sarma, B., Santiso, E.E., Simpson, J., Fry, J.C., Yurttas, N., Biermann, K.L., Chen, J., Trout, B.L., Myerson, A.S., 2016. On the connection between nonmonotonic taste behavior and molecular conformation in solution: the case of rebaudioside-A. J. Chem. Phys. 143 (24). 244301-1 to 244301-10 (Chapter 5, Section 3).

Claridge, T.D.W., 2016. High-Resolution NMR Techniques in Organic Chemistry, third ed. Elsevier, Amsterdam, ISBN: 0080999867. 538 pp.

Cox, J.R., Simpson, J.H., Swager, T.M., 2013. Photoalignment layers for liquid crystals from the di-π-methane rearrangement. J. Am. Chem. Soc. 135, 640–643 (Chapter 5, Section 1).

Edgar, S., Zhou, K., Qiao, K., King, J.R., Simpson, J.H., Stephanopoulos, G., 2016. Mechanistic insights into taxadiene epoxidation by taxadiene-5α-hydroxylase. ACS Chem. Biol. 11 (2), 460–469 (Chapter 4, Section 2).

Harris, R.K., 1986. Nuclear Magnetic Resonance Spectroscopy: A Physicochemical View. Longman, Essex, England, ISBN: 0582446538. 282 pp.

Jacobsen, N.E., 2017. NMR Data Interpretation Explained: Understanding 1D and 2D NMR Spectra of Organic Compounds and Natural Products. John Wiley & Sons, Hoboken, ISBN: 1118370228. 640 pp.

Jacobsen, N.E., 2007. NMR Spectroscopy Explained: Simplified Theory, Applications and Examples for Organic Chemistry and Structural Biology. Wiley-Interscience, Hoboken, ISBN: 0471730963. 688 pp.

Liu, X.-L., Lipp, J.S., Simpson, J.H., Lin, Y.-S., Summons, R.E., Hinrichs, K.-U., 2012. Mono- and dihydroxyl glycerol dibiphytanyl glycerol tetraethers in marine sediments: identification of both core and intact polar lipid forms". Geochim. et. Cosmochim. Acta 89, 102–115 (Chapter 5, Section 4).

Pretsch, E., Bühlmann, P., Badertscher, M., 2009. Structure Determination of Organic Compounds: Tables of Spectral Data. Springer, New York, ISBN: 3540938095. 433 pp.

Index

Note: Page numbers followed by *f* indicate figures and *t* indicate tables.

Printed in the United States
By Bookmasters